HITLER'S GIRL

HITLER'S GIRL

THE BRITISH ARISTOCRACY AND
THE THIRD REICH ON THE EVE OF WWII

LAUREN
YOUNG

HARPER ⬤ PERENNIAL

NEW YORK • LONDON • TORONTO • SYDNEY • NEW DELHI • AUCKLAND

HARPER ● PERENNIAL

A hardcover edition of this book was published
in 2022 by HarperCollins Publishers.

HarperCollins books may be purchased for educational, business,
or sales promotional use. For information, please email the
Special Markets Department at SPsales@harpercollins.com.

FIRST HARPER PERENNIAL EDITION PUBLISHED 2023.

Designed by Nancy Singer

Page ii, Union Jack: Philip Openshaw/Shutterstock
Page iii, Reichsadler Deutschland: @Engineer/stock.adobe.com

Library of Congress Cataloging-in-Publication Data has been applied for.

ISBN 978-0-06-293674-5 (pbk.)

23 24 25 26 27 LBC 5 4 3 2 1

To
Anna Sophia, Charlotte, and Paul

CONTENTS

HITLER'S GIRL

PROLOGUE

If you pluck a chicken one feather at a time,
nobody notices.
—*Mussolini*

I wrote most of this book unthinkably isolated with my family in our New York City apartment during a global pandemic. Our solitude was punctuated by a terrifying blare of ambulances speeding up and down the avenue, their sirens amplified by the empty streets. There was nowhere to go and a lot of time to reconsider many things.

The political theater in the press was one of our few distractions, with daily accounts of brazen assaults on our democratic institutions, both at home and abroad. I had the nagging sense that the postwar liberal democratic heritage that we had always assumed would be our birthright was starting to feel like a losing hand of poker. The game keeps us engaged, although the winning hands are few—and there seems to be a lot of cheating. As defenders of the democratic tradition, we were watching the chips pile up in adversarial hands, hoping that the tide would change, without ever folding and reevaluating both the gravity of our loss and a better strategy.

As I write, eighteen states in the United States have passed laws making it harder for citizens to vote. After decades of efforts by the Federalist Society, the composition of the US Supreme Court has endowed seats for life to an ultraconservative majority that does not reflect the will of the people on banner issues like abortion. Political polarization, which has marched hand in hand with income inequality since the end of the Second World War, is reaching new vertiginous heights, amplified by a cohort of vaccinated Republican governors who wave the flag of civil liberties in relentless maskless anti-vax campaigns that are filling hospitals to capacity with unvaccinated believers dying of COVID. While we can protest the infringement on our civil liberties to end a global pandemic through vaccine mandates, we are free to eclipse a woman's basic sovereignty over her own body exemplified by the restrictive abortion law recently passed in the state of Texas, taking direct aim at *Roe v. Wade*, a Supreme Court decision dating to the 1970s that is supported by a majority of Americans. The tides of anti-Semitism are rising, not just in the United States where reported acts of anti-Semitism remain at historic highs. Alarmingly, basic truth, either spoken by our elected officials or in the echo chambers of social media, is elastic and negotiable.

The losing chips seem to be piling up.

None of the assaults on our democracy are new, just the examples we use to describe them. There has been a debate about the democratic continuum that goes back to the sages of ancient Greece. In fact, these symptoms of democratic decay are all too familiar. After all, Hitler was democratically elected and even voted for by many Jews in the vain hope that being on his side would spare them the worst of his threats. German Jewish newspapers in the 1930s flatly refused to believe that Hitler would execute on his worst stated instincts. Europe in the 1930s had many similar warnings to today, mostly ignored or rationalized.

A new wave of populism, both in Europe and the United States, demands that we view the rise and ultimate demise of Fascism in the 1930s differently. Sometimes, these lessons are found in the least likely of places. What started as an ordinary country pub lunch on an unusually sunny Sunday before the Brexit vote in June 2016, just days before the British vote to leave the EU, revealed the enduring British fascination with the legacy of World War II. Billboards proclaiming "Gerries Go Home!" evoked the Second World War front and center at the core of the British psyche today. Remarkably, the archives of British history present an entirely different and hidden history of Britain in the 1930s, one marked by German complicity at the highest ranks of British society. The narrative of the brave forces upholding democracy and a fragile new world order could just as easily have tipped the other way in England, the flash point of liberal democracy in the days leading up to the war.

There are many implicit parallels to England in the 1930s in today's world. One of the most troubling is complacency and its repercussions. During this period in England, from Chamberlain's policy of appeasement to a groundswell of support for Hitler among the British ruling class, the future of democracy turned on the head of a pin. The potential for Britain to succumb to Nazi Germany was a real possibility—and a probability that was avidly advanced by an influential segment of the country's elite and systematically covered up by its government as too dangerous to reveal, even decades later. The statute of limitations on many intelligence files classified during this era, some of which were categorized as such as recently as during the Kennedy administration, are expiring now at another urgent moment. The irony of this information becoming available at a similarly perilous ideological moment presents a great opportunity to take a closer look at this period, which until now has never been examined from this perspective.

The rise of Fascism in Nazi Germany was powered by the humiliating loss of the First World War and those whose interests were overlooked by twenty successive and unsuccessful Weimar governments leading up to Hitler's election to chancellor in 1933. Fascism in England was arguably even more insidious, largely the handiwork of the elite ruling class in a bid to preserve its power. Influential segments of the British aristocracy formed a murky fifth column to Nazi Germany, and elements of the British ruling class almost succeeded in tipping the pendulum in Hitler's favor. How could the twenty-two-year-old aristocrat Unity Mitford meet with Hitler more than 160 times between February 1935 and September 1939, espousing Nazi vitriol, without the British government ever taking a real interest? A dangerous combination of complacency and complicity among those with power and influence in England during the 1930s nearly toppled a proud and long-standing democratic tradition.

If Britain, along with the rest of Europe, had fallen to the Nazis, approximately a quarter of the world's population would have lived under Fascist control, likely for generations. While the story of the Nazi march through continental Europe is well documented, Britain's ideological vulnerability has been largely overlooked. The British archives also reveal the peril with which democratic institutions were challenged in England during this period and tap into contemporary debates about immigration, religious differences, income and gender inequality, free speech, the press, and the moral obligations of ethical political leadership. How far those in power were willing to go to maintain their grip is the untold tale of Britain in the 1930s.

. . .

I began to think about the period of the 1930s in England some years before the accelerated erosion of democracy that has characterized the maelstrom of political life in the United States, Britain, and many other European countries since. My interest originated in my classroom at the London School of Economics, where I taught about the policy of appeasement. This happened at the same time as the statute of limitations on intelligence files expired on many classified documents locked up for posterity. I was immediately struck by the irony of these documents' being declassified at a time when our own Western liberal democratic tradition was facing serious challenges. As I began digging through these files at the British National Archives at Kew, what I found was not the defense of the policy of appeasement, which I expected, but a hidden history of collaboration and a newfound sense of alarm.

I also found that my access to further significant information was curtailed. Clearly, there are many who do not want these stories to come to light today. The archives at Chatsworth House, which house Unity Mitford's diaries given to her sister, Deborah Mitford Devonshire, after Unity's death, are closed indefinitely. The Royal Archives at Windsor have been notoriously closed to researchers interested in specific correspondence between the English royal family and their German cousins during this period, especially the correspondence between Queen Mary and her cousin the Duke of Saxe-Coburg and Gotha, who was sent by Hitler to recruit the royal family to the German cause. There has also been speculation that the correspondence between Queen Mary and her cousin was so inflammatory that most of it was stolen by Anthony Blunt, the Keeper of Royal Portraits at Windsor Castle and a double agent for Russia as part of the Cambridge Five spy ring. In fact, most of the Cambridge Five was actually trained by Guy Liddell, the intelligence officer who waited on the docks at Folkestone Harbor but was prevented from questioning Unity Mitford directly when she was repatriated to England after her extended time in Germany. I found myself wondering whether this was the context that prompted Kim Philby, one of Burgess's trainees, to summon his own privilege as the perfect cover for treachery. A tape of Philby speaking to East German spies in 1981 and later recovered in Stasi files, notes him remarking that he "got away" with his misdeeds "because I was upper class."

Despite all of the controversy around Blunt, he eventually became the dean of the Courtauld Institute in London, the esteemed academy for art historians. While so many personal papers that hold the keys to many unanswered questions about this era are impossible to access, Blunt's personal papers are still held at the Courtauld Library. It did not escape my notice that while both Chatsworth House and the Royal Archives were both essentially closed to me, I could peruse Blunt's earliest notebooks from his first trips to Russia, ostensibly when he was recruited by the Russian intelligence services. I could only conclude that the topic of double-crossing spies paled by comparison to whatever was being held at Chatsworth House and at Windsor.

It is possible that the stories these archives might reveal are presumably perceived to be too dangerous and there are still people who make it their mission to protect them. Yet there is plenty available, and the more I researched, the more I concluded that the history of this era was not only one of appeasement but also one of active collaboration that demonstrated over and over how far the ruling aristocratic class would go to protect their position. The British royal family's German origins—the Battenbergs changed their name to Windsor in 1914—and the ongoing upstairs/

downstairs of British society during this time, divided between a small and insular aristocratic elite and their subjects, is well-trodden territory. However, the implications of this class-based society and the threats posed to British democracy have been largely overlooked. There were many more supporters of Hitler among this small ruling elite than previously thought, a fact that was actively covered up. At the same time, Hitler was quite strategic in recognizing the value of Pan-European aristocratic alliances to advance the Nazi cause. He recruited many German aristocrats, most especially those with well-placed English cousins, such as the Duke of Saxe-Coburg and Gotha, a German-born grandson of Queen Victoria and first cousin of Queen Mary of England. Members of the English aristocracy, also the leaders of government, were often active and enthusiastic supporters of Nazi Germany and were in a position to create and advance government policies to protect their own power.

While historians are increasingly looking at the rise of Fascism in Eastern Europe, Germany, and Italy during the 1930s as a road map to understanding the perils of today's world, England in the 1930s has been largely overlooked. Much of this has to do with secrecy. Most of the evidence raised in this book has been classified and reclassified by successive British governments, deemed too explosive in its exposure of a time when British democracy was in real jeopardy.

Alarmingly, the rise of nationalism and populism, both in the United States and Europe today, bear many of the same hallmarks. At what point does complacency become complicity? Will democracy require a similarly cataclysmic event like World War II in order to reassess and ensure its survival? The maelstrom of pandemic, racial inequality, economic turmoil, and climate change provides the cover of chaos for a similar power grab. In order for democracy to continue to thrive as the cornerstone of Western governance, we must understand its vulnerabilities and from them safeguard and adapt democratic institutions for the citizens of the twenty-first century.

• • •

What if? What if this dangerous inflection point in England in the 1930s had tipped toward Fascism instead of democracy? Was the British ruling class united behind the British government during the period of the 1930s, or did they have an entirely different agenda?

CHAPTER 1

GERMAN FASCISM CROSSES
THE ENGLISH CHANNEL

D rawing back the curtain on the insidious and largely suppressed history of the Far Right in England during the 1930s necessitates a stop in Germany, which nominally exported Fascist ideology to England. The rise of the Nazi Party in Germany finds its roots in the legacy of the Great War. In November 1918, having lost the support of the military and with his government collapsing, Kaiser Wilhelm II fled Germany for Holland, never to return. In the intervening years until Hitler became chancellor in 1933, successive coalition governments in Weimar Germany created a weak and ultimately vulnerable political landscape. The longest continuous government managed to stay in power for just two years. The economic ruin propagated by the punishing terms of the Treaty of Versailles as a factor in the collapse of successive

Weimar governments is the most familiar, but not the only factor contributing to the rise of the Nazi Party in Germany during this period.

The legacy of the Great War, where over two and a half million Germans died and four million were wounded, formed a discernible line in the sand between the victors and the vanquished. Citizens on both sides were convinced that the sacrifices to support the war effort, including the obliteration of an entire generation of young men in the trenches of Europe, were worth making. The governments of the winning axis congratulated themselves that in victory, their war aims had been achieved and life returned to its regular prewar rhythm, albeit without sons, husbands, and brothers.

For Germany, the vanquished, losing the war left the turmoil of economic uncertainty, unbridled violence, and boiling resentment. The element of violence was exacerbated by the expanded arsenal of lethal weaponry introduced to the battlefield: mortars, hand grenades, flamethrowers, and mustard gas. These weapons created mass casualties, which, in the case of German soldiers, were expeditiously triaged at rudimentary hospitals at the front so that they could return to the trenches to fight on. As a result of hasty and inadequate treatment of injuries, legions of maimed soldiers returned home after the war as cripples, often to the streets with few job prospects, due to both their physical condition and the devastated Weimar economy.

The struggle for survival was consigned not only to former soldiers but to the generation of women left behind to support their hungry families. One notable feature of the Weimar economy was the many brothels, well regulated and subject to medical oversight, which sprang up all over the country, even contiguous to the fighting trenches. The brothels were serviced by desperate women of all ages, including elderly grandmothers, all in search of a way to support their families in the face of destitution.

In addition to increasingly punitive laws enacted against any perceived threats to the accelerating power of the Nazi machine, the decade of the 1930s in Germany made it very clear who belonged and who didn't. In 1937, the work of the German expressionists, including Marc Chagall, Wassily Kandinsky, Paul Klee, Emil Nolde, George Grosz, and Ernst Ludwig Kirchner, was exhibited in Munich in a large show known as *Entartete Kunst* or *Degenerate Art*. This exhibition encompassed images of the reality of the streets of the Weimar Republic: the crippled soldiers, aging and desperate prostitutes, the thin veneer of the cabaret society. Desperation was painted against the backdrop of the industrialization that was sweeping Europe, despite the dire economic environment in Germany. Spectators were eager to see it. Over the course of several months about two million people, more than twenty thousand visitors per day, viewed the works, often waiting in long lines in freezing weather.

These distorted images of the underbelly of Weimar society were juxtaposed in the popular imagination by nationalistic evocations of German folkloric ideals—blond country maidens and sturdy Hitler youth. Art became more explicitly anti-Semitic. Details of Jewish facial features were exhibited at the Deutsches

Museum (Munich) in an exhibition called *The Eternal Jew*, advertised by a poster featuring a grotesque caricature of a Jewish man evoking a rat. During a limited run of three months, from November 1937 through January 1938, this exhibition attracted more than five thousand visitors per day.

The legacy of the First World War, captured by German degenerate art, was also marked by a culture of violence. Returning soldiers, seeking to reclaim their past military glory, took to the streets and formed marauding bands of paramilitary groups that aggressively and lawlessly prowled German cities. They continued to wear uniforms, took part in parades, won medals, and relived the brotherhood of wartime. Prolonging the soldier life created a sense of empowerment, an attempt to dull the eviscerating sense of having lost a generation-defining conflict. While veterans of the Great War joined these paramilitary groups, younger German men who were too young to fight, and with few job prospects in the postwar economy, eagerly participated, too. The sense of elitism created by membership in paramilitary groups inevitably established obvious outsiders who did not, could not, or chose not to belong. Uniformed and armed groups of returning soldiers now fought on the streets of Weimar Germany, nominally a battle against the rising tide of Communist agitators, but also for the sake of fighting. Steeped in the battlefields of untold destruction and death, the paramilitary groups transferred methods and practices of trench warfare to the streets, adding a violent undercurrent to the complicated and unstable postwar politics of the Weimar Republic.

One of the early paramilitary groups, the Reichsbanner, was formed in opposition to the KPD (Kommunistische Partei Deutschlands), or Communist Party. The Reichsbanner, which by the beginning of the 1930s had over three million members, of which 250,000 were youth members, was eventually banned by the Nazi Party in an effort to consolidate the Far Right into a single political entity and eliminate competing groups. Founding Reichsbanner members ultimately were imprisoned in concentration camps. While banning these groups, Hitler also exploited the culture of violence among returning soldiers and unified them under the Nazi mantle.

The processes of "brutalization" brought by the paramilitary groups of Weimar Germany resulted not only in street violence, but in the mass killing of civilians as well as prominent German politicians and members of the Communist Party. Groups such as the Freikorps, the largest and most deadly of the paramilitaries, murdered three hundred suspected Communists in Mitau, in the Balkans, as well as suspected Communist agitators like Rosa Luxemburg and Karl Liebknecht, all of which was overlooked and tolerated by the Nazis when they came to power.

The paramilitary groups also introduced and normalized political violence, which became a hallmark of the Nazi regime as it fortified its grip on Germany during the 1930s. As early as November 1921, the Freikorps assassinated Reich finance minister Matthias Erzberger in what was a second assassination attempt. Erzberger was regarded by the militia as a traitor to the people and a symbol of the maligned Weimar government. He was murdered while on vacation, walking through the Black Forest, by two former soldiers who fled Germany and were later granted amnesty by the Nazi Party.

The following year, in June 1922, foreign minister and Jewish industrialist Walther Rathenau was also assassinated by members of the Organizational Council of the Freikorps. Rathenau, who built his family's electrical engineering company, AEG, was also a friend of Albert Einstein, who later fled Nazi Germany. As foreign minister, Rathenau was a marked man by the right-wing militias for several reasons. First, he insisted on paying reparations owed under the terms of the Treaty of Versailles, a deeply unpopular stance. Rathenau also negotiated the Treaty of Rapallo with the Soviet Union in 1922. This treaty was viewed as uniting Germany with the pariah Communist state and encouraging Communist elements in the country.

After Hitler was democratically elected chancellor in January 1933, the chaos and violence of the Weimar years provided cover for his ascent. The Reichstag fire in February 1933, just a month after Hitler was sworn in as chancellor, was quickly and conveniently blamed on a Dutch Communist agitator, but was likely perpetrated by the Nazis as a pretense to defend the country. Ironically, Hitler's democratic election facilitated his rapid consolidation of power. The following month, the Enabling Act, signed into law by President Paul von Hindenburg, as the last gasp of the Weimar Republic, gave Hitler broad powers to bypass the government. This was the same year that Hitler's memoir and manifesto, *Mein Kampf*, was published in Britain.

While the streets of German cities like Berlin and Munich

were marked by violence and disability, the streets of London were deceptively calm. Creeping right-wing tension also simmered in England, for different reasons, and out of the public eye. With an entire generation of young men lost to the Great War, Britain struggled to return to life as it was before. But they did, with the scars of loss and the commitment to avoid another war at any cost. The war also represented a turning point in Britain's colonial ambitions. The sun was starting to set on the glory days of the British Empire. By 1913, the British Empire comprised more than 412 million people, and by 1920, about 24 percent of the earth's population. The First World War, however, began to reveal the strains of maintaining a far-flung empire and the cost, both in loss of human life and economic drain, in fighting a "total war."

The threats to the British upper classes, from the loss of life of the Great War, the existential threat of Communism, and the burgeoning cracks in the British Empire, also produced a subtle shift in an effort by the aristocratic elite to preserve their way of life. Images of the young Queen Elizabeth giving the Nazi salute in mid-1933 were published by the *Sun*, Britain's largest circulating tabloid, in July 2015. The future queen, about seven years old when the picture was taken, is shown with the Queen Mother and Elizabeth's three-year-old sister, Margaret, in a still photo taken from a seventeen-second grainy family film made at Balmoral Castle. Edward, Duke of Windsor, the former king Edward VIII, stands in the background, ostensibly coaxing the young girls. When the image surfaced, a spokesman for Buckingham Palace responded that "it is disappointing that film, shot eight decades ago and apparently from (Her Majesty's) personal family archive, has been obtained and exploited in this manner." Another royal household source claimed that the publication had "exploited" the Queen and commented in defense of the royal family that "most people will see these pictures in their proper context and time. This is a family playing and momentarily referencing a gesture many would have seen from contemporary news reels." Nonetheless, despite this being an image of children at play, it unavoidably documents a link between the British royal family and Nazi Germany.

This film is just one artifact that illuminates the Fascist incursion that was creeping into British politics in the early 1930s. Popular history has tended to attribute this movement exclusively to the British Union of Fascists (BUF). Oswald Mosley, the charismatic leader of the BUF, initially known as the New Party, announced his party platform with much fanfare on February 28, 1931. Despite fighting pneumonia, Mosley made a rousing speech and claimed the role of leader of the British Far Right. A profile in the *Observer* published in January 1933 immediately made a comparison to Hitler, observing that "where Mosley is so like Hitler is in his sense of the dramatic. There is an extraordinary atmosphere of drama about a Mosley meeting, a sense that great things are about to happen."

Mosley himself was a controversial character, a marginal player in British political life with a network of aristocratic friends and a paradoxically broad appeal to the working-class dockworkers of East London, ironically also home to London's Jewish community. One observer remarked when she met Mosley in 1923 that "so much perfection argues rottenness somewhere." Membership in the BUF grew quickly in its first two years, aided by the support of the press magnate Lord Rothermere, owner of the largest circulating broadsheet, the *Daily Mail*, which publicized the successes of the party. In January 1934, the *Daily Mail* ran a widely read editorial, penned by Rothermere himself, "Hurrah for the Blackshirts," in which he refers to storm trooper–inspired uniforms worn by BUF supporters. Rothermere praised Mosley for his "sound, commonsense, Conservative doctrine." Rothermere further touted the BUF as "the organized effort of the younger generation to break this stranglehold, which senile politicians have so long maintained on our public affairs." Other widely read newspapers similarly promoted the BUF, running headlines like "Give the Blackshirts a Helping Hand" and "Blackshirts Will Stop the War." An editorial in the *Spectator* refuted Rothermere's support for Mosley, writing that "the Blackshirts, like *The Daily Mail*, appeal to people unaccustomed to thinking. The average *Daily Mail* reader is a potential Blackshirt ready made."

By 1934, there were about fifty thousand members of the BUF. To this day, most records of MI5 or Foreign Office intercepts of the workings of English Fascist groups are still classified. Similarly, attempts have been made to obscure the sources of the early funding of the New Party, as it was first known, before it became more established as the leading right-wing party in England in 1933. While Mosley contributed a substantial share of his personal fortune, he also called on an extensive circle of friends to donate, a list that included some of the most illustrious aristocrats, politicians, and heads of business in England. Rumors swirled that Lord Inchcape, the Scottish shipping heir, was an early backer in addition to Lord Nuffield, founder of Morris Motors and later Nuffield Health. Nuffield believed that Mosley was "a real leader" and "the one bright spot on the political horizon."

Mosley also used his good friend, the socialite Sibyl Colefax, and Lady Houston to tap their fellow high society friends. Harold Nicolson, an early BUF supporter, husband of Vita Sackville-West, and later an MP who served in Churchill's government, famously retorted, "[A]lthough I loathe antisemitism, I do dislike Jews." Nicolson wrote in his Londoner column of the *Evening Standard* in April 1931 that the New Party headquarters were staffed by "young men with fine foreheads," dashing "from room to room carrying proof-sheets and manifestos" among the tapping of type-writers and ringing telephones. Other MPs associated with the New Party/BUF included Viscount Lymington, Henry Drummond Wolff, Sir Patrick Hannon, Lord Mount Temple, Lord Brabazon of Tara, and the Marquis of Clydesdale. The BUF also had foreign backers. By 1933, Mussolini was sending significant funds that arrived by a circuitous route via Switzerland, and disbursed by the Italian ambassador to England, Count Dino Grandi, in small denominations, an action vigorously denied by the home secretary when it was raised at a cabinet meeting during the war.

These early and influential donations enabled the BUF to move headquarters from Belgravia at 12 Grosvenor Place overlooking the gardens of Buckingham Palace, to a much larger location, the "Fascist Fort" as it was nicknamed, in Chelsea on the Kings Road.

Another supporter recalled that by 1933, the Blackshirt HQ or the "Black House" as it also came to be known, was "filled with students eager to learn about this new, exciting crusade; its club rooms rang with laughter and song of men who felt that the advent of Fascism had made life worth living again." BUF membership was inexpensive—1 shilling per month if employed, a fourpence a month if unemployed—and at a time of high unemployment, there was work available at the BUF headquarters, which offered accommodations as well as a salary of £1 to £2 per week.

Mosley's movement attracted the interest of conservative leaders of both business and politics who were critical of the slow pace of economic recovery in England. One increasingly central aspect of the BUF platform was anti-Semitism, which evolved in its virulence as the decade progressed. Increasing public rancor also likely contributed to the party's eventual decline. The *Evening Standard*, a popular broadsheet, ran a series of articles in 1936 about the way the East London Jewish community was being targeted by BUF supporters. "Police cars patrol slowly by, and under the lamplight stand police horses. . . . Casting a wider circle still one finds small parties of Blackshirts parading the main streets, bickering here and there in a childish fashion with Jews . . . shouting slogans that spell the letters of Sir Oswald Mosley's name and occasionally singing the unpleasant song with the refrain 'We've got to get rid of the Yids.'"

Jewish refugees from Hitler's Germany became another flashpoint political issue for the BUF. The MP for North Tottenham made a direct call to end the "invasion of undesirable aliens,"

Jewish refugees fleeing the recently announced Nuremberg laws (1935) in Nazi Germany. In an assessment of anti-Semitism in Britain in 1935, the Jewish Labour Workers' Circle wrote that "Mosley seizes his opportunity to play upon the old racial and religious prejudices. . . . 'The Jews are to blame!' he shouts. 'The Foreigners are the cause of your misery!' And in their wretchedness and bitterness some of the victimized workers imagine that they see prosperity amongst the 'foreigners.' They think that the Jew is not so badly off as they themselves are, and the vital struggle for social emancipation and freedom becomes side-tracked into a 'national' and 'racial' issue. As the situation becomes worse, and as the class-struggle becomes keener, the ruling class makes full and terrible use, through its agents, of the unspeakable weapon of anti-Semitism."

Sir Anthony Rumbold, whose father was ambassador to Germany from 1928 to 1933, tried to alert the British government about the rise of the Nazi Party and the treachery of Mosley's BUF, which borrowed from Nazi German Fascism an ideology with which he was well acquainted. Rumbold described going to a BUF meeting with "the Mitford girls" on April 22, 1934, at the Royal Albert Hall, one of the most prestigious concert halls in central London. Rumbold and the Mitfords sat in a box while an orchestra played "England Awake," lyrics set to the Nazi hymn the Horst Wessel Song. "Everybody in the box saluted, except me, I was embarrassed and slunk away."

Rumbold, however, was the exception. The BUF continued to expand its meetings to larger venues. Their biggest rally to date was held at the Olympia, a large exhibition hall in an upscale neighborhood of leafy green streets and large homes in West London, on June 7, 1934. While BUF rallies were held in gentrified areas of West London, the rank-and-file BUF foot soldiers mostly hailed from the rough-and-tumble slums of East London. Approximately ten thousand BUF loyalists attended the rally at Olympia in addition to about five hundred protesters, who yelled, "Hitler and Mosley, what are they for? Thuggery, buggery, hunger, and war." The author Nancy Mitford recalled that Mosley "brought a few Neanderthal men along with him and they fell tooth and (literally) nail on anyone who shifted or coughed. One man complained afterward that the fascists' nails had pierced his head to the skull." The violent methods used by the BUF to evict the protesters shocked so many that the party saw a decline in membership after the rally. Lord Rothermere also rescinded his support after the violent clashes.

Unity Mitford, daughter of Lord Redesdale and a member of the inner circles of both Mosley and Hitler, disagreed with criticism of the BUF. She wrote to her mother, "Aren't the English newspapers absolutely nauseating. All this absurd and lying outcry about 'brutality' at Olympia only goes to show the urgent necessity for getting rid of disgusting Jewish influence in all walks of English life. If the Jews can imagine that they can, with the help of a few insignificant Tory M.P.'s, manufacture lies which will kill our movement, I'm afraid they're quite wrong. Of course exactly the same thing happened in Germany in the early days of the NSDAP (Nazi Party)."

While Mosley's BUF was noisy and sometimes even violent, the undercurrents of British Fascism were more widespread and nefarious, infiltrated by Hitler at the highest echelons of British society. Hitler actively pursued an effective and largely overlooked backdoor foreign policy with a less traditional spin. Suspicious of his own Foreign Ministry, he recruited German aristocrats with close family ties across Europe. After the destruction of the First World War, the royal families of Britain, France, Belgium, Netherlands, Spain, Italy, Romania, and Germany were all looking to seize a more visible political role in their respective countries. Hitler understood early on the strength of this transnational elite, united and emboldened by a fear of Communism and desperate to avoid the same violent end as their cousins, the Russian czar and his family. As Queen Marie of Romania wistfully remarked, "Fascism, although also a tyranny, leaves scope for progress, beauty, art, literature, home and social life, manners, cleanliness, whilst Bolshevism is the levelling of everything." In the case of the British aristocracy, fears of Communism were compounded by the first tremors of the unraveling of the British Empire—and their place at its vaulted helm.

As he consolidated his power after his election as chancellor in 1933, Hitler capitalized on these fears and exploited this politically powerful cross-border network to advance his own national socialist platform, with stunning results—especially in England. The British ruling class's interest in Fascism was first cultivated by visits to Italy in the 1920s, predating Hitler's ascent. Members of the British upper classes such as Harold Nicolson, the Duke of

Westminster, and the Duke of Buccleuch all traveled to Italy to observe the new politics of Fascism, where the term originated. Winston Churchill visited Italy as well during this period and left with a positive impression of this new political movement. "Italy has shown that there is a way of fighting the subversive forces. . . . She has provided the necessary antidote to the Russian poison. Hereafter no great nation will be unprovided with an ultimate means against the cancerous growth of Bolshevism."

Hitler used German aristocrats to call on their English cousins. In April 1933, Otto von Bismarck, a name that still resonated in England for his grandfather's role in the unification of Germany, officiated at a debate at Chatham House. Colonel Christie, a member of the British security services who was present, noted that "the fact that a man of Count Bismarck's breeding and tradition has given his wholehearted support to the Nazi Movement should persuade us to examine without prejudice the underlying principles of this somewhat feverish nationalism which has been accepted by millions of well-educated Germans." In March 1934, Bismarck reported back to Germany that the Duke [of Connaught] is very interested in Germany, acknowledging his useful position of influence within the British royal family. The Duke of Connaught was not the only British aristocrat targeted. The Earl of Kincardine and his brother Lord Ronald Graham wrote to the German embassy about their interest in German labor camps. They requested to "see if possible a Labour Service Camp and a concentration camp" in order to understand how the Nazis effectuated "race purity and fitness." By 1934, there were documented executions in German concentration camps, the first steps of ethnic cleansing by the Nazi regime.

Bismarck also established a close relationship with the Duke and Duchess of Windsor, the former Edward VIII and Wallis Simpson. Bismarck and his Swedish-born wife, Ann-Mari, would help the unmarried German ambassador Ribbentrop in London to entertain the Windsors, mostly in Marbella, where they both had homes. This was the beginning of a lifelong association.

Hitler's initial forays into the sphere of influence of the British royal family were at first fairly straightforward. In 1934, he approached Princess Victoria Louise, the daughter of the former kaiser, Wilhelm II, and a Nazi sympathizer. She wrote in her memoir that they "received an astounding demand from Hitler, conveyed to us by Ribbentrop. It was no more nor less than we should arrange a marriage between (our daughter) Friedericke and the Prince of Wales. My husband and I were shattered. Something like this had never entered our minds, not even for a reconciliation with England. . . . We told Hitler that in our opinion the great difference in age between the Prince of Wales and Friedericke alone precluded such a project, and that we were not prepared to put any such pressure on our daughter." Hitler actively capitalized on any opportunity he saw to create alliances between German and British aristocrats, although in this case, a protective mother spoiled the plan.

Hitler's most reliable and well-placed German connection to the British nobility was Carl Eduard, Duke of Saxe-Coburg and Gotha, brother of Princess Alice of England and grandson of

Queen Victoria, who as a child spent a great deal of time at the English court. The German spy Carl August Clodius revealed during interrogations by the British in 1946 that "in pursuit of an Anglo-German rapprochement (the Duke) offered his social connections and tried to invite to Germany many prominent Englishmen and put them in touch with important people" there. Princess Alice wrote in her memoirs that even though her late husband, Lord Athlone, chancellor of the University of London, often objected to German ambassador Ribbentrop's descriptions of the "New Deutschland," it did not preclude him from accepting a donation of *Germaniae Historica* in 1937 with much fanfare at the university.

After a long absence, Coburg's trips to England accelerated during the early 1930s, with holiday visits to Sandringham in 1932 and 1933 to stay with George V and Queen Mary. After that, he traveled to England several times a year, usually coinciding with important political events.

Coburg used his sister's country estate, Brantridge Park, as a base from which to conduct his shadow diplomacy on behalf of Hitler. In 1936, the secretary of state for war and close friend of the Duke of Windsor, Duff Cooper, understood that Princess Alice's weekend invitation was politically motivated: "The point of it was to meet the Duke of Coburg, her brother. It was a gloomy little party—so like a German bourgeois household. . . . I was tactfully left alone with the Duke of Coburg after luncheon in order that he might explain to me the present situation in Germany and assure me of Hitler's pacific intentions."

The intelligence community was also keeping tabs on these

unofficial overtures to well-placed British relatives. In a report that Anthony Blunt passed on to his handlers in Moscow, he noted that Ribbentrop's senior staff shuttled between Berlin and London, and "their job, in essence, was to influence the broadest possible range of British public opinion in a pro-German Direction. (Ribbentrop's office) thus included individuals with connections in royal as well as diplomatic, political and industrial circles."

In his memoir, *Inside the Third Reich*, Albert Speer, the architect of many of the monumental stadiums and buildings that were constructed as testament to the rising strength of the Nazi vision, wrote of Hitler's strategic view of Britain when he rose to chancellor. Speer did not believe that Hitler's initial aim was war with Britain. He recalled a comment Hitler had made in Obersalzburg: "I really don't know what I should do. It is a terribly difficult decision. I would by far prefer to join the English. But how often in history the English have proved perfidious. If I go with them then everything is over for good between Italy and us. Afterward, the English will drop me and we will sit between two stools."

The perception of pro-German sympathies espoused by the British aristocracy was established in a keynote speech given by the Prince of Wales, the future Edward VIII, to the British Legion on June 11, 1935. In the speech, the prince praised a forthcoming trip to Germany by a delegation of British servicemen. The trip consisted of a two-hour audience with Hitler, a visit to Dachau in which the prison guards dressed as prisoners to camouflage what was actually happening there, a visit to a British war cemetery, as well as "a quiet family supper with Herr Himmler." Anthony Eden, minister of foreign affairs, deemed this visit a German propaganda stunt.

Chips Cannon, an American-born member of Parliament, married into the Guinness family and therefore also related to the Mitfords by marriage, remarked that the prince's speech resulted in "much gossip about the Prince of Wales' Nazi leanings," likely influenced by his wife's introduction to the German ambassador von Ribbentrop. Cannon observed that "he (the Prince) has made an extraordinary speech to the British Legion advocating friendship with Germany; it is only a gesture, but a gesture that may be taken seriously in Germany and elsewhere." Basil Newton, a member of the staff of the British embassy in Berlin, noted that Goering gave a speech in Germany soon after supporting warmer relations between the two countries and specifically referring to "Germany's pleasure at the recent declaration of His Royal Highness the Prince of Wales. The German ex-servicemen and the German nation cheerfully grasped the hand which had been stretched out at them."

The archives reveal that Hitler himself cultivated the Nazi sympathies of the British upper class, who actively pursued a Fascist agenda. But to what end?

CHAPTER 2

MORE DANGEROUS THAN
THE BLACKSHIRTS

Secret Right-Wing Movements in Britain

With Germany tugging at the sleeve of many in England, Odette Keun, a Dutch-Turkish émigré writer, reflected on the perception of Germany in Britain, likely also influenced by her lover, H. G. Wells. The essay, "Perfidious Albion," published on July 6, 1935, observed that "there was no consciousness of Nazism as a reality in the public mind. The English man in the street is ignorant; the English intellectuals are apathetic; the English popular press is dumb." Keun's observation characterized an apathy that allowed the emergence of other Fascist movements in England, notable for recruiting the rich and powerful—and keeping it a secret. The inaugural meeting of the Anglo-German Fellowship, whose membership included fifty sitting members of Parliament, was attended by Carl Eduard, Duke of Saxe-Coburg and Gotha, first cousin of the Duke of Windsor.

While history remembers Oswald Mosley and the British Union of Fascists, there is a darker tale of Fascism among other notoriously more exclusive and secret right-wing groups. Among the earliest of these groups was the Link, founded in 1937 by Admiral Barry Domvile. Domvile was preoccupied by a conspiracy he believed was created by an association between Jews and the Freemasons, which he called "Judmas." He cautioned that Britain had fallen into the orbit of the United States and was being manipulated by Jewish financiers, whose influence only Hitler was managing to hold at bay. Link members included influential patrons like Lord Redesdale, a member of the House of Lords and Unity Mitford's father; Archibald Ramsay; and the Dukes of Westminster, Marlborough, Wellington, Buccleuch, Bedford, and Hamilton. Unlike some of the more rarefied and exclusive right-wing groups, Domvile also aggressively sought to expand membership among rank-and-file Brits across the country, recruiting new members in smaller British cities such as Birmingham and Southend. Touring the outer branches of the Link chapters, Domvile remarked that "everywhere I go I am asked about the Jews and what we are going to do about them."

Domvile was later imprisoned for treason under Defence Regulation 18B, the Internment of Enemy Civilians Act, and the club was disbanded. It was briefly revived by Ian Fleming, then a naval intelligence officer and later the author of the James Bond 007 novels, and Maxwell Knight, head of MI5, as a trap to attract Rudolf Hess, Hitler's second-in-command, to England. Unknown to the Nazi high command, Hess parachuted into Scotland in May 1941, seeking to track down the German sympathizers in the British government that he had heard so much about and with whom he was determined to negotiate a peace treaty. Instead, he was imprisoned and later tried at Nuremberg for crimes against humanity. He committed suicide in prison in 1987.

While it has long been assumed that Hitler's racism was cultivated directly by German culture, steeped in the nationalist music of Wagner and the Teutonic myth of strapping blond youth, Hitler was fascinated by what he believed was endemic American racism, especially its long and established history of racial discrimination of minorities. As Hitler formulated his racial policies, he often looked to what he believed was an American model. He was fascinated that slavery was written into the American constitution, along with calls to "exterminate" Native Americans. Apparently amused by claims of equality in America, he cited the immigration acts of the 1920s, which were eventually used to prevent Jews fleeing Nazi Germany from finding refuge in America. When Roosevelt convened an international conference on the plight of Jewish refugees in Evian in July 1938, the German Foreign Office responded that they found it "astounding" that the United States would criticize Germany's treatment of Jews and then refuse to allow them into their country.

Hitler was also an admirer of American eugenics and studied California's sterilization laws, an ongoing program last revised in

1917 and continued to be practiced into the 1960s in which more than twenty thousand sterilizations were performed on mentally ill people, most belonging to racial minorities. This California program inspired the Nazi sterilization law of 1934, also known as the Law for the Prevention of Offspring with Hereditary Diseases, which gave the Nazis the right to sterilize any citizen they believed would jeopardize the purity of Nazi lineage and the precursor to the horrific sterilization experiments conducted in the concentration camps.

Zyklon B, the gas of choice used in the gas chambers of Nazi death camps, also found roots in America. The gas was first used to execute an inmate in Nevada in 1924, when it was established that it required a gas chamber in order for it to be effective and for the gas not to escape. Zyklon B was used for executions in the United States until 1999, when a California court ruled that it was cruel and unusual punishment. The use of a gas chamber and Zyklon B was revived in Arizona as recently as June 2021 for the execution of death row prisoners due to a shortage of lethal injections, another repercussion of supply chain shortfalls. American execution chambers were later equipped with chutes into which poisonous Zyklon pellets were launched. The American inventor of this device explained that "pulling a lever to kill a man is hard work. Pouring acid down a tube is easier on the nerves, more like watering flowers."

The Nazi executioners took note and built a similar device, enabling more lives to be taken more quickly. During the Nazi era, Zyklon B was manufactured by IG Farben, located next to

Auschwitz, and licensed to American Cyanamid. In addition to Zyklon B, this particular IG Farben facility used approximately 83,000 slave laborers during the span of the war to manufacture Zyklon B and synthetic rubber to support the Nazi war effort. IG Farben was also the target of Allied reconnaissance missions in April 1944. With the gas chambers of Auschwitz billowing smoke in the photos taken by South African reconnaissance pilots, the Allied goal was not to liberate Auschwitz but to undermine wartime production at the IG Farben plant. These photos are now held at Yad Vashem, the Israeli Holocaust memorial in Jerusalem.

The sphere of influences linking Hitler to American precedents, from Jim Crow laws to Zyklon B, also found an incarnation in British right-wing movements. The Nordic League, first known as the White Knights of Britain, or the Hooded Men, and said to have been founded by two Nazi spies in London, described itself as "an association of race conscious Britons [and being of the service] of those patriotic bodies known to be engaged in exposing and frustrating the Jewish stranglehold on our Nordic realm." The reference to both the White Knights and Hooded Men conjured up the American Ku Klux Klan, which dated to the state of Tennessee in 1865.

The Nordic League was led by Archibald Ramsay, a conservative MP, later also the founder of the Right Club. Ramsay, who frequently quoted *The Protocols of the Elders of Zion*, a fabricated volume of claims accusing Jews of seeking world domination, was a decorated war hero and graduate of Eton and Sandhurst. Members of the Nordic League were known only by numbers and used secret passwords like "Perish Judah" to gain entry to meetings where Nazi marching songs and anti-Jewish German propaganda films were sprinkled into the agenda. MI5 infiltrated meetings of the Nordic League and assessed MP Ramsay to be generally "unbalanced," but not a threat worth prosecuting as the Nordic League did not reach a broad population of British citizenry. Members of the early Nordic League included many of the principal members of what was later to become the secretive Right Club: the Duke of Wellington, William Joyce (Lord Haw-Haw), A. K. Chesterton, Lord Brocket, the Duke of Hamilton, and Lady Douglas-Hamilton. In Nazi Germany, the Nordic League was viewed as "the British branch of Nazism."

Ramsay was also a frequent contributor to the *Anglo-German Review*, a magazine published between 1936 and 1939 and funded in part by German ambassador Ribbentrop's London embassy funds, which sought to promote good relations between England and the rising Nazi regime. The *Anglo-German Review* offered a glossy view of Germany to interested British readers. It advertised a trip organized by Modern Touring of Regent Street that promised "Eighteen Days through Germany, Austria and Czechoslovakia for £31," as well as swastika badges, a German production of the British opera *Maria Stuart* at the Battersea Town Hall, and a touring group concert by German accordion players who played in London and other cities. German events aimed at enticing English patrons to Germany were also advertised in the British press. These included the Cologne Dog Show, a dance competition in Kassel, as well as the Nazi-approved exhibition of Degenerate Art in Stuttgart. A returning reader, a Rhineland holiday maker, and Link Club member, praised the "physical perfection" of German children, their blond braids and "high spirits." One popular column in the *Anglo-German Review* was aimed at British housewives, praising the example of their German counterparts: "meticulous in every detail appertaining to home comfort, the German housewife sees that her kitchen is efficiently equipped before she begins to think of her personal needs. A fur coat can wait."

In 1939, Ramsay sought to extend membership for the Fascist cause that was gaining momentum among the political and social elite and founded the Right Club. The formal membership of the Right Club was a carefully guarded secret, but the list with Ramsay's lengthy annotations of the club's notoriously well-connected members was recorded in a red, leather-bound ledger that came to be known as the Red Book. The Red Book was considered so sensitive that it was kept out of the public view until 1990, when, with little fanfare, it was mysteriously donated to the Wiener Holocaust Library collection in London, the world's oldest and largest repository of Holocaust archives and collections relating to the Nazi era. While the members of the Right Club were some of the most prominent fixtures of British society, there has been little published about its activities or membership. At his later internment trial, Ramsay argued that the membership list was protected because he believed that the police and intelligence services would conspire with Jewish interests to personally undermine the members. Ramsay also specifically disavowed the Nazi regime and said he generally disagreed with their ideologies, with the exception of their policy toward "the disruptive activities of Organized Jewry."

The stated objective of the Right Club "was to oppose and expose the activities of Organized Jewry . . . (and) to clear the Conservative Party of Jewish influence." While their mission statement did not refer specifically to the policy of appeasement, the club's other stated purpose was "to avert war which we considered to be mainly the work of Jewish intrigue centered in New York."

Members of the Right Club wore a small badge in their lapels, an eagle killing a snake with the initials "PJ" or "Perish Judah." The membership met irregularly and the organization was maintained by voluntary donations ranging from £3 to £100 per annum. The Right Club operated out of quiet and inconspicuous locations in South Kensington, including a flat belonging to Margaret Bothamley at 67 Cromwell Road, where frequent parties were held. Bothamley was also an active member of the Nordic League, the Link, and the British Union of Fascists. Meetings would begin with a hymn such as "The Land of Dope and Jewry," written by Ramsay on a piece of stationery from the House of Commons:

> Land of dope and Jewry,
> Land that once was free,
> All the Jew boys praise thee
> While they plunder thee.
> Poorer still and poorer
> Grow the true born sons,
> Faster still and faster

They're sent to feed the guns.
Land of Jewish finance,
Fooled by Jewish lies,
In press and books and movies
While our birthright dies.
Long still and longer
Is the rope they get
But, by God of battles
T'will serve to hang them yet.

As a decorated member of the British military, Ramsay successfully recruited a younger generation of officers to the club's membership, including Prince Yuri Galitzine, whose high-society wartime wedding was closely covered by the press, and Lord Ronald Graham, an upcoming naval officer and fixture on the London party circuit. An index in the back of the *Red Book* indicates ancillary Fascist societies with whom the Right Club membership was attempting to coordinate, including the National Socialist Party, the Imperial Fascist League, the Nationalists, and the Nordic League. A Scottish chapter of the Right Club based in Edinburgh was also launched and suggested the increasingly mainstream reach of the Far Right in England. The Right Club Edinburgh branch was presided over by a steering committee that included representatives of the Rotary Club, the Business Club, and Edinburgh University, as well as Robert Bell, a member of the Scottish Parliament.

The annotated list of Right Club members located in Central

London included John Milner Bailey, later the 2nd Baronet, who was married to Churchill's daughter Diana. Another member, William Baker, was listed as the founder of the Economic Reform Club, which later became the renowned Economic and Social Research Council. Other aristocratic members included Princess Evelyn Blücher von Wahlstatt, born in Lancashire and married to a German prince; Lord Charles Carnegie, 11th Earl of Southesk, married to Princess Maud of Fife, a granddaughter of Edward VII, who Ramsay commented was "a very loyal and patriotic man," publicly espousing right-wing views until his death in 1992. The 12th Earl of Galloway served as a warden of the club, as did the Duke of Montrose, who was eventually barred from the House of Lords for his extremist views. Lord David Redesdale, a member of the House of Lords and Unity Mitford's father, was also a member of the Link and the Anglo-German Fellowship. He recalled meeting Hitler, a "right-thinking man of irreproachable sincerity and honesty." The Duke of Wellington, who belonged to several other right-wing groups, also served as a warden.

Among Ramsay's annotations in the *Red Book* was to highlight Commander E. H. Cole, a chancellor of the White Knights of Hooded Men, whom he flagged for supporting a statute enacted by Edward I in 1275 that called for tight restrictions on landownership and moneylending by Jews in Britain. Ramsay attempted to raise this statute again as one of his last acts at Parliament. William Joyce (Lord Haw-Haw the humbug of Hamburg) left England for Germany at the start of the war, where he joined fellow Right Club member Dorothy Erskine as a broadcast journalist—of German propaganda. One British journalist observed that Joyce was "thin, pale, intense, he had not been speaking for many minutes before we were electrified by this man . . . so terrifying in its dynamic force, so vituperative, so vitriolic." In spite of this, a BBC poll revealed his show to be quite popular in Britain.

Church leaders also belonged to the Right Club. The unitarian minister Reverend George Coverdale-Sharpe was also a member of the Nordic League and the Link. He attacked the Bishop of Durham for having "the audacity to say that our Lord was a Jew." Other prominent members of the Right Club included Frances "Fay" Taylor, an Irish race car driver, also known as "Flying Fay of Dublin," later interned under Regulation 18B.

The Right Club was infiltrated by MI5 operatives Kim Philby and Guy Burgess, both secret agents for Russia and two of the Cambridge Five. As a result of their espionage on behalf of Britain, some of the most prominent members of the Right Club were eventually imprisoned, under Section 18B, in Royal Holloway Prison (Mosley, Ramsay), the Isle of Man, Brixton, and other British prisons. These scions of British politics and high society found themselves in jail, shoulder to shoulder with petty thieves and sex offenders. Diana Mitford famously wore her furs in Royal Holloway, while her husband, Oswald Mosley, having complained about his lack of comfort in prison, was allowed to employ the services of a petty thief as his valet.

The Right Club was debated in Parliament, deliberations that were reported back to Berlin by the new German ambassador to London, Herbert von Dirksen, who had replaced Ribbentrop in

May 1938. In a cable, von Dirksen reported encouragingly that parliamentary debates about Britain receiving Jewish refugees from Nazi Germany showed that "an appreciation of the Jewish question is also on the increase in Britain . . . reflected by the suppression of all reports of Mosley's Fascist meetings, which are sometimes very well attended."

Debates such as those in Parliament about Britain's role in accepting Jewish refugees from Germany were a rallying cry for Ramsay and the Right Club, which opposed relaxing Britain's immigration laws, citing anti-Semitic tropes as justification. One member, Arnold Spencer Leese, published a newsletter, the *Fascist*, headquartered on Craven Street in Bloomsbury. Leese's publication called for the extermination of Jews even before the Nazi death camps became known to the general public. Leese was also eventually imprisoned under Section 18B.

While right-wing groups with German-leaning sympathies like the Right Club and the Nordic League gained influence in England during this period, engagement with the Nazi regime became more mainstream. Even the Boy Scouts sought contacts with Nazi Germany. Lord Baden-Powell, founder of the Boy Scouts, had a meeting with the German ambassador in November 1937, at which Hartmann Lauterbacher, chief of staff to the Hitler Youth, was in attendance. Baden-Powell said that meeting "opened my eyes to the feeling of your country towards Britain which I may say reciprocates exactly the feeling which I have for Germany" and recounted that the Nazi high command recommended that he visit Hitler in Germany after a planned trip to Africa. Baden-

Powell confirmed that "Ribbentrop seemed very much in earnest and was a charming man to talk to, has many relations in England and I knew his uncle in India." Determined to forge a "peaceful" alliance with Germany through the nations' youth, Baden-Powell conducted joint exercises between the Boy Scouts and the Hitler Youth, mostly in England, in 1936–37. Organized by Baden-Powell, a brigade of Hitler Youth, dressed in full uniform, performed Nazi songs at a church in Dalston in 1937.

Right-wing groups and their influential members also propelled official British foreign policy toward Germany during this period. Chamberlain's signature policy of appeasement, also known as the "Peace Party," advocated for a political agenda that had as its principal aim to avoid another world war at nearly any cost. This included allowing Germany to flex its expansionist inclinations on the Continent so that Britain could continue to rebuild its military—and enable its empire to flourish. One of Chamberlain's first acts when he was named prime minister in May 1937 was to conclude the Munich Agreement, which effectively turned a blind eye to the German annexation of the Sudetenland, an ethnic German area of the former Czechoslovakia. While this agreement was celebrated by Lord Redesdale, the Duke of Wellington, and Sir Ernest Bennett, a member of Parliament, as a victory, it also theoretically aligned Britain with the Nazi Party against the Communist threat.

In an article in the *Anglo-German Review*, the trio praised "the joint efforts of Herr Hitler and Mr. Chamberlain" for finding a policy that righted the wrongs brought by "the men of Versailles," referring to the punishing terms of the treaty. They added that "the vast majority of our countrymen trust the pledge of both the Fuhrer and our own good Prime Minister." In September 1938 Neville Chamberlain visited Berchtesgaden, Hitler's home in the German Alps, with George VI's knowledge and support, to negotiate the Munich Agreement, which both men regarded as a great success for their policy of appeasement.

Lord Redesdale used the conclusion of the Munich Agreement as an opportunity to praise Hitler during a debate in the House of Lords: "[T]he gratitude of Europe and of the whole world is due to Herr Hitler for averting a possible catastrophe of such magnitude (i.e., civil war in Austria) without shedding one drop of blood." Another supporter in the ensuing parliamentary debate said that through the Munich Agreement, Chamberlain had "snatch(ed) from the very jaws of war the prospect of peace." The entire Link Club signed a letter to the editor of the *Times* in 1938 praising the Munich Agreement for redressing the political and economic damage done to Germany through the Versailles Treaty. The argument ran even deeper in its appeal to common sense: Germans and Britons had much in common. Both wanted peace. Eden eventually resigned as foreign secretary in protest of the Munich Agreement. His change in position and eventual resistance to Chamberlain's appeasement policy were feared to be leading Britain closer to war. It was not until 1942 that Churchill himself dissolved the British Peace Party. By then, Britain had been at war with Germany for over two years.

While Chamberlain's policies, the Munich Agreement chief

among them, eventually led to a humiliation on the world stage, he found enthusiastic support for his views from a group known as the Cliveden Set, perceived to be so powerful that the British popular press characterized them as a shadow government. The Cliveden Set gathered at the baronial surround of Cliveden House, the stately home outside of London belonging to Lord and Lady Astor. Located just five miles from Windsor Castle and fewer than thirty miles from the halls of power at Westminster, it was easily accessed by a wide range of the rich and powerful, who were welcomed as (mostly) weekend guests. Perched on the top of a crest, overlooking the Thames valley, the house dates to the Stuart kings. This wasn't the first or the last time that Cliveden played an important supporting role in British politics, a distinction that dated back to the Hanoverian dynasty and later liberal firebrands in the nineteenth century.

Cliveden House was a wedding gift from William Astor to his son, Waldorf, later Viscount Astor and a politician in his own right, to mark his marriage to the American-born Nancy Lancaster. Like many of the great houses of England during the First World War, Cliveden was converted to a field hospital. Soldiers were brought from the trenches of France and Belgium, limbs disfigured by the new, advanced weaponry of the Great War and, more often than not, also blinded by mustard gas.

The *Observer* and the *Times*, both newspapers owned by Waldorf Astor, were early and forceful sympathizers of Hitler's Germany. The *Observer* portrayed Hitler as being "definitely Christian in his ideals," praising him for both "moderation and common sense," while the *Observer* called the rising number of violent Nazi brawls to be "a manifestation of organizing power, of discipline, of earnestness and of confidence." It also encouraged its readers not to assess "too sinister an interpretation" to the Nazi success in the 1933 election and quick consolidation of power.

The Astors were also visible and voluble supporters of Chamberlain's policy of appeasement. Portrayed in the press as "the most important supporters of German influence here," wielding a position of extraordinary political power, they became known as the leaders of the Cliveden Set. When Ribbentrop was named ambassador to England, he courted the Astors, gaining an invitation to spend the weekend at their home in Sandwich in the Kent countryside near Dover. They reported staying up until late at night discussing Hitler's plan to create a nexus of power between Germany and England. The group, often led by Philip Kerr, 11th Marquess of Lothian, appointed ambassador to the United States in 1939, believed that the League of Nations was unable to confront the rising problems in Europe and that British foreign policy must focus on finding a peaceful arrangement with Germany as a platform with which to attract other European allies, notably France.

The Astors were hosts of great renown. One lunch party at Cliveden, on October 24, 1937, became the catalyst for something entirely new, a "queer Anglo American gathering" that for years has "exercised so powerful an influence on the course of British policy." Among the guests that day were Foreign Secretary Anthony Eden and his wife; Nevile Henderson, ambassador to

Germany; Alec Cadogan, soon to be appointed undersecretary of the Foreign Office; Geoffrey Dawson, editor of the *Times*, as well as a few others.

The press linked that weekend at Cliveden House and Nancy Astor in particular to Eden's resignation as well as his replacement with the more politically pliant Lord Halifax. As the recently appointed foreign secretary, one of Halifax's first trips was to Berlin in November 1937, ostensibly to visit a Nazi-organized hunting exhibition (Halifax was master of the Middleton hounds), a convenient pretext for a meeting with Hitler. Lord Londonderry, a frequent guest at Cliveden, has been credited with suggesting to Halifax that he meet with Hitler as quickly as possible. Halifax advanced this idea of a trip to Germany to Churchill just ten days before his planned departure, at an invitation by Goering.

Halifax's presence in Germany was a stroke of good luck for the Nazis. As the car wound its way up the mountain and eventually stopped in front of an imposing gate, Halifax recounted that "as I looked out of the car window, on eye level, I saw in the middle of this swept path a pair of black trouser legs, finishing up in silk socks and pumps. I assumed this was a footman who had come down to help me out of the car and up the steps, and was proceeding in leisurely fashion to get myself out of the car when I heard von Neurath throwing a hoarse whisper at my ear of '*Der Fuhrer, der Fuhrer.*'" Halifax found Hitler to be "most sincere" and Goering "frankly attractive." The *Evening Standard* reported that Hitler told Halifax that Germany would offer Britain a ten-year truce if they allowed Germany a free hand in Central Europe.

Chamberlain proclaimed Halifax's visit "a great success," and even the incredulous Eden found himself in agreement.

Increasing chatter about the shadow government that operated under thin disguise of weekend parties at Cliveden House or "Schloss Cliveden," increased to a louder pitch by the mid-1930s, with claims that the Astors were operating "a junta" and "a cabal," a group so powerful that they formed "God's Truth Ltd." Others referred to them as "a deadly secret committee." In June 1936, the *Week* published a piece claiming that the Cliveden Set, which also met occasionally at the Astors' London residence in St. James's, had reached an "extraordinary position of concentrated political power (becoming) one of the most important supports of German influence."

On any given weekend, the rich and the powerful gathered, hosted by the Astors and united by their common belief that British foreign policy needed to pivot toward a rising Germany to protect their own influential place in the new world order. The guest list might include the prime minister, George Bernard Shaw, T. E. Lawrence (of Arabia), the governor of the Bank of England, the Aga Khan—and even Charles Lindbergh. The Aga Khan, Eton-educated and a British citizen, acted as Hitler's contact with the British government during the 1930s. He later offered to raise thirty thousand Arab troops in the Middle East to support Hitler's military operations in the region. While the Aga Khan's family denies these claims, recently declassified cabinet files confirm his role and debate the gravity of this threat to the balance of power in the region and British interests in Palestine. The *Washington*

Post characterized the "Cliveden Set" as "the real center of British foreign policy, challenging the constitutional structure of British Democracy." The common link among all the guests was a sympathy toward Hitler and his Reich, and the promotion of a British foreign policy that promoted both.

Nancy Astor was portrayed as a central figure in the power wielded by the Cliveden Set, making her a target for the British press, which disparaged her because of her pro-German politics. She was portrayed in cartoons pulling the strings of government puppets from her perch at Cliveden and as a gargoyle, along with a duo of houseguests, in a series in the *Evening Standard* dubbed "The Shiver Sisters." Its caption was "Any Sort of Peace at Any Sort of Price." Another cartoon shows her dancing in a tutu with Goebbels, conducting music called "German Foreign Policy." One accuser claimed that "the conduct of the pro-Nazi element in our National life as represented by yourself (Nancy Astor), various members of your family and certain aristocrats who have apparently dictated our foreign policy since Chamberlain took over . . . [you] should shamefacedly retire from public life."

As an American expatriate, Nancy Astor had a reputation as the leader of an influential pro-German lobby at the heart of the British government, who fearlessly criticized "the appalling anti-German propaganda here (in England)," which also opened her to criticism in the United States. The head of the American Jewish Congress Women's Division wrote in a letter to "Dear Lady Astor" that "if the Jews in America are against Nazi Germany, it is because they conceive it to be their duty as Americans to battle for civilization and humanity and therefore to stand against the crimes of Hitlerism." Nancy responded erroneously that the Zionist movement was founded at Cliveden House. There were no links between Cliveden House and the British founders of the Zionist movement. Nancy was also accused of anti-Semitic politics in the House of Lords, covered widely by the British press as evidence of the danger of the pro-Nazi agenda she advanced. Nancy's defense was categorical, stating bluntly that criticism of her politics was "a false and stupid story published in a Communist rag." George Bernard Shaw, a frequent guest at Cliveden, proclaimed that "never has a more senseless fable got into the headlines."

By 1938, the Cliveden Set, not just Nancy Astor, was a frequent topic in the British press. There are allegations that the Germanophile leanings of the group were largely an invention of Claud Cockburn, whose journalistic coups about the Cliveden Set were published weekly in the *Week*. Evidence suggests that Cockburn's coverage revealed uncomfortable facts about the sway of a small group of influential people in British political life during the period of the 1930s. Cockburn was not alone in his coverage. The *Reynolds News*, a pro–Labour Party Sunday newspaper, ran a feature story about the Cliveden Set describing them as a group of British aristocrats and other influential British leaders, notably press barons, endeavoring to draw British foreign policy closer to Germany.

By 1938, as German aggression in the Sudetenland became widely anticipated, Chamberlain spent a late March weekend at Cliveden House. In between games of charades, the decision was made to allow Germany to annex Czechoslovakia, a region of little strategic importance to Britain and therefore an event not worthy of starting a second world war. Lord Lothian, named ambassador to Washington a few months later, in August 1938, summed up the decision in support of Chamberlain: "He is the only person who steadfastly refused to accept the view that Hitler & the Nazis were incorrigible & would understand nothing but the thick stick." From Chicago, Lothian wrote to Nancy Astor that "the Cliveden Set yarn is still going strong everywhere here," bluntly adding that "it symbolizes the impressive spread by the left and acceptable to the average American that aristocrats and financiers are selling out democracy in Spain and Czechoslovakia because they want to preserve their own properties and privileges." Anthony Eden, whose political demise was linked directly to the Cliveden Set, proclaimed "how terrible has been the influence of the Cliveden Set," while Harold Nicolson, who famously stood up against the Munich Agreement, claimed that they were "a defeatist pampered group."

The last meeting of the Cliveden Set took place on the eve of the war during the summer of 1939. Their agenda had been overturned by the inevitability of war. Ribbentrop, now the German foreign secretary, embarked on an early June weekend in 1939 on one of the last missions on behalf of the German government in England. He sent as an emissary Adam von Trott zu Solz, a friend of Waldorf Astor from his Oxford days. Unknown to his hosts, he was sent in an unofficial capacity by the German Foreign Ministry to probe the group to see whether there was any way to avert

England's entry into the war. Von Trott was in good company for such an inquiry. Among the other thirty guests that weekend were the chief supporters of a pro-German agenda for British policy: Lords Lothian and Halifax; Geoffrey Dawson, editor of the *Times* and old friend of Lord Halifax; and Sir Thomas Inskip, secretary of state for the Dominions.

Despite the political sway of those that gathered at Cliveden that June weekend, it was too late for von Trott's proposals. England threatened to declare war if Germany again invaded another sovereign nation. Chamberlain made a radio broadcast from the cabinet room at 10 Downing Street at 11:15 A.M. on September 3, 1939, fifteen minutes after Germany failed to respond to an ultimatum about the invasion of Poland: "This country is at war." In a five-minute speech, he continued that "you can imagine what a bitter blow it is to me that all my long struggle to win peace has failed. Yet I cannot believe that there is anything more, or anything different that I could have done and that would have been more successful. Up to the very last it would have been quite possible to have arranged a peaceful and honourable settlement between Germany and Poland. But Hitler would not have it." To which Nancy Astor replied, "I can't believe that has happened." Chamberlain died just a year later, in November 1940.

Cliveden House remained a place of political intrigue even after the war. An epilogue to the Cliveden Set was the Profumo Affair, which largely unraveled at Cliveden during the early 1950s. Young John Profumo resigned in scandal as secretary of war in 1963 over an affair with the nineteen-year-old Christine Keeler, who was rumored to have contacted intelligence operatives in the Soviet government. A recently declassified MI5 file written in 1940 reveals that prior to the Keeler affair, Profumo also had a twenty-plus-year relationship with a Nazi spy, Gisela Klein, known to him as Gisela Winegard. Klein ran a disinformation service in Paris on behalf of the Nazis. She was also rumored to have had a baby with a high-ranking Nazi official. Even Nancy Astor apparently warned Profumo that Klein was potentially a spy. In 1951, she used Profumo as a reference in a visa application to travel with her American husband from Tangiers to London.

How might the Profumo Affair have been received if it had been revealed that Christine Keeler was not the first woman to prey on Profumo's political position, but that he maintained a long-standing relationship with a Nazi spy? Profumo's intelligence file did not rule out that Profumo knew that Klein was a Nazi spy. Additionally, this relationship, well known to British intelligence, did not hamper Profumo's rise to power in the British government—until another similar scandal brought him down.

CHAPTER 3

A SHORT HISTORY
OF THE LONG HISTORY
OF BRITISH ANTI-SEMITISM

As the British Union of Fascists and other right-wing groups gained momentum in England, the dignified world of British politics was challenged by increasingly bold anti-Semitic politics and violent clashes with its opponents. This was not an entirely new development. A simmering undercurrent of anti-Semitism had been present in Britain going back at least to the turn of the century, and probably longer.

Jews in Britain had a long history. The Board of Deputies of British Jews was founded in 1760 by Sephardic Jews, and by 1810 expanded to include Ashkenazi Jews from Central and Eastern Europe. But starting in the late nineteenth century with rising tides of anti-Semitism in England as well as the increasingly frequent

massacres of the Russian pogroms, British Jews sought a home-
land, a place safe from persecution. British Jews arrived in the
port of Jaffa in the late 1890s, before the First Zionist Conference,
chaired by Theodor Herzl in Basel, Switzerland, in August 1897.
Tel Aviv was founded in 1909 by sixty-six Jewish families, many of
them wealthy British Jews, who used seashells to divide the sandy
and desolate tract of land that was soon to become the country's
capital into homesteads for each family.

By the early 1930s, the England that these families left behind
was increasingly hostile to the Jews who lived there, approximately 300,000 people, or 0.65 percent of the population. Adver-
tisements for job vacancies included an addendum that Jews were
not "admissible." Insurance companies boycotted Jews as a bad
risk, one company stating that Jews were "untrustworthy individuals." In May 1932 a Leeds synagogue was vandalized. At the Leeds
Crown Court, a trial was held over a fatal car accident caused by
a Jewish driver. The coroner exclaimed that there were too many
Jews on the jury for a fair trial to be held. In December 1933, the
Middlesbrough Motor Club banned Jews from membership, an-
nouncing that it was "of the opinion that Jews and Gentiles do not
mix socially in numbers." In Glasgow, Jews were excluded from
certain government-built apartments while the Scottish undersecretary proposed quotas for Jews entering the country, initially
targeting Jewish émigrés escaping from Germany.

The political activists and German Jewish émigrés Mathilde
Wurm and Dora Fabian were found dead in a London flat they
shared on Great Ormond Street in April 1935. The cause of death

was determined to be Veronal poisoning due to "unsound mind." Because there was no motive established, it was assumed by their friends that they were murdered by the Gestapo—in Central London. Veronal, a barbiturate that was easily obtainable without a prescription, was the drug that was also speculated to have been used in a later suicide attempt by Unity Mitford in Munich in September 1939. It was also notable that when pressed about the murders of Wurm and Fabian, the British government declined to make any statement about the case.

The British film industry, in particular, was targeted with anti-Jewish slurs by powerful and respected critics. The Norwich MP excoriated Jewish control of the British cinema, claiming that "there are millions of boys and girls in this country . . . (whose) souls are being taken from them as blood money for a syndicate of dirty American Jews." Canon Palmer of the Ilford Catholic church delivered sermons about the "evils of (the) unclean film industry," suggesting in a homily that a "strict boycott of all picture houses should be considered until Jewish filth is swept right away."

Although acknowledged to be a falsification, *The Protocols of the Elders of Zion* was translated and still fairly widely circulated in England by the mid-1930s. It was part of a toxic brew of other anti-Semitic rhetoric circulating, such as the *G.K.'s Weekly*, penned by G. K. Chesterton, the writer and philosopher who depicted Jews as taking advantage of Christian values, and the *National Review*, which portrayed the Jews as a fifth column expediting fears of an imminent invasion by Germany. Notices also started to appear on lampposts and in public areas with the slogans "Christ Was the

First Fascist" and "Any Jew Is Worth Two Englishmen." Gateshead, near Newcastle, was home to a religious Jewish community dating to the mid-nineteenth century. The Gateshead Bensham Synagogue was desecrated twice during a three-week period in 1935, papered with posters proclaiming "Down with Jews."

Germans also exported German culture to England. During the period of the 1930s, the Nazi government attempted to infiltrate England by supporting British institutions. One example is an edition of *Mein Kampf* published in England in 1939 with proceeds designated to benefit the British Red Cross. *Mein Kampf* was first published in England in 1933. While unsuccessful initially, *Mein Kampf* eventually sold an average of three thousand copies per year in England as compared to ten thousand per year in the United States, across a much larger population.

Other pro-Nazi publications were also supported and advanced by those in power. Future prime minister Harold Macmillan sponsored the 1940 publication of *Unfinished Victory* by leading British historian Arthur Bryant, a work that advanced the Nazi platform and was specifically anti-Semitic. The book was published by Macmillan and Co. with Harold Macmillan acting as Bryant's personal contact and representative. Bryant was also recommended by Foreign Minister Lord Halifax and sent to Germany by Neville Chamberlain to represent Britain.

Anti-Semitic acts were frequently recorded by Jewish newspapers in Britain during the decade of the 1930s—quickly rationalized and almost never prosecuted. The Jewish Labour Council in Britain issued an assessment of anti-Semitism in 1935 that they mostly attributed to German influence on the activities of the BUF. They summarized that "the example of Hitlerism has provided them [the BUF] with a lead in the vilest forms of Jew baiting and anti-Jewish hysteria, and no pains have been spared in a deliberate attempt to convince the British public that the Jews are responsible for all the evils of the day." Far from dominating the wealth of the country, they point out, many Jews in Britain actually live below the poverty line. The article goes on to query "what is the purpose of this racial hatred" in England? "This blind racial prejudice serves a special purpose for the ruling class, who use it as a means of diverting the wrath and discontent of the workers from itself against some convenient helpless minority. The ruling class hires speakers and writers who readily prostitute whatever abilities they have in order to delude and smother the just protests of the suffering workers. And what more convenient scapegoat can they find than the mass of Jewish workers, who live in all countries as an unprotected minority? (There are 350,000 Jews in Britain, less than 1% of the total population.) This was the policy of the Russian Czars in maintaining their rule by organizing pogroms as a 'safety valve' for every kind of unrest and political crisis."

A special parliamentary debate on a wave of anti Semitic incidents across London was called in July 1936. The MP for Hammersmith argued that if the government did not act, there would be "pogroms in this country." Synagogues were vandalized in Leeds, Gateshead, Manchester, and many parts of London. The Gateshead synagogue alone was desecrated two or three times per week in 1935. A pig's head was left at a synagogue in Bethnal

Green during Passover 1939, and boycotts of Jewish-owned shops were urged across the country.

The cornerstone of the Chamberlain government's policy of appeasement was to avoid entering war at any cost. Foreign Minister Lord Halifax stated at a cabinet meeting in November 1938, just after Kristallnacht in Germany, that "his anxiety was that Germany should not go to war with us as a consequence of any action we might take to help the German Jews." This included a policy to close Palestine to German and Austrian Jewish refugees fleeing the Nazi regime. Halifax declared that he had "reached the conclusion that perhaps the best course was to do nothing as any positive action on our part would only make the position of the German Jews still worse." The cabinet agreed that it was a mistake to muddy the diplomatic waters "with this Jewish question."

The British government imposed strict quotas on Jewish immigration to British-controlled Palestine. US ambassador Joseph Kennedy and Neville Chamberlain embarked on a plan to ship Jewish refugees to Africa instead. The later Kindertransport to Britain, the resettlement of children fleeing Nazi Germany, was not nearly as robust a response to saving Jewish children as it has been portrayed. One clause required that potential adoptive families have expensive insurance policies. Additionally, the British government did not officially recognize these children as immigrants. Their status in Britain was only temporary and stipulated that they come to Britain without their parents, essentially creating orphans. Had the Kindertransport been less onerous, more Jewish children undoubtedly would have been saved.

CHAPTER 4

THE DEBUTANTE NAZI

While Hitler was building bridges to the English ruling class, the young aristocrat Unity Mitford was firmly entrenched in life in Munich, transfixed by her quest to meet Hitler. Unity Valkyrie Mitford was born in 1914 in Swastika in Ontario, Canada, where her parents were prospecting for gold to offset the family debts. The fourth daughter, she was one of the famed Mitford sisters who would become legendary fixtures of London high society. Her middle name was derived from her grandfather's love of Wagnerian opera.

Given Unity's later renown as a rabid Nazi, both her name and birthplace were an ironic precursor to her future. Unity met with Hitler more than 160 times during a five-year period leading up to the outbreak of the Second World War with full knowledge of the British government and intelligence services. She was determined to captivate Hitler and she succeeded, as chronicled in the diary of Eva Braun, who after a jealous suicide attempt at the height of Unity's powers of persuasion over Hitler, eventually married him (and died with him) in the final days of the Reich.

The Mitford family was not alone in their declining family fortune. In the period between the wars, British aristocratic families began to sell off their family estates and often-empty London homes. Lord Montagu wrote in *More Equal than Others* that after the First World War, one quarter of the land in England had changed hands, representing the biggest transfer of land in the country since the Norman Conquest. After moving back from Canada, having gained a respite from their mounting debts, the Mitford children grew up in eccentric and relatively unbridled isolation at Swinbrook House in the Cotswolds, raised by nannies, largely ignored by their parents, and speaking a secret language called Boudeledidge.

Like other girls of her social class, Unity made her society debut in May 1932, curtsying before the king and queen at Buckingham Palace. At six feet tall, Unity was described by her sister Deborah as a "huge and rather alarming debutante . . . (who) towered over her fellows at various debutante functions rather like a big Santa Claus among the Christmas dolls." Her cousin Anita Leslie dubbed Unity "the least handsome of an outstandingly beautiful tribe." Unity's introduction to society also coincided with a move from the family's country home, Swinbrook, to Old Mill Cottage about thirty miles from central London. The Court

Circular of the *Times* of London breathlessly recounted the guest lists of weddings, dances, and dinner parties that a young debutante like Unity would have attended, as well as a benefit for the Hampstead League of Mercy. Diana Mitford, then newly married to Bryan Guinness, heir to a family brewing fortune, held a ball for Unity and another friend on July 7, 1932.

Unity, however, did not embrace the role of a debutante. Her younger sister, Deborah, recalls that she "shone like an enormous peacock in flashing sham jewels." She also wore her pet snake, Enid, around her neck as jewelry, stole stationery from Buckingham Palace for friends, and was rumored to let loose her pet rat at society balls, to liven up the party. Her cousin Anita further noted her "obstinate refusal to flirt with eligible gentlemen"—all of which contributed to the unlikely outcome that she would finish her introduction to British society with a marriage proposal, as was typically expected of a young girl of her social class.

While Mosley's BUF party was essentially working class from the slums along the docks of East London, Mosley himself had strong ties to the British aristocracy, and ultimately directly to the Mitford family. Both Unity Mitford and her sister Diana were quick to join the BUF. Mosley's first marriage in 1920 was to Cynthia ("Cimmie") Curzon, daughter of Lord Curzon, the former viceroy of India. Diana, married at the time to Bryan Guinness and the mother of two small children, became Mosley's mistress in 1932. Mosley was not in a rush to leave his wife, although her unexpected death in May 1933 from peritonitis, a month before

Diana's divorce from Bryan Guinness was finalized, presented the opportunity. Diana accelerated a scandal that her parents were eager to avoid.

To steer clear of unseemly gossip surrounding Cimmie's death, Mosley encouraged Diana to travel during the summer of 1933 and take Unity with her to Bavaria, a place where she had never been. The trip had been proposed by the chance meeting of Ernst "Putzi" Hanfstaengl at a dinner party in the London home of Mrs. Richard Guinness, a cousin of Diana's former husband. Putzi offered to introduce Diana to Hitler if she decided to go to Munich, an offer she quickly accepted. Since she had become "as thick as thieves" with her younger sister, Unity, she took Unity along with her. Unity seemed like a natural travel companion, given her interest in the BUF. Unity and Diana viewed the trip to Germany as an opportunity to see for themselves the first initiatives of Germany's new leader, whom both sisters, as members of the British Fascist Party, had admired from afar.

German-born and Harvard-educated, Hanfstaengl, Hitler's foreign press secretary, held court at some of the most distinguished dinner parties in London. Hanfstaengl was visiting England with his friend William Randolph Hearst. As he recalled, "I was on good terms with the Guinnesses, and met Unity through them in London—she was very much second string to her sister, tagging along. Otto von Bismarck contacted me as foreign press chief, with the request to introduce the two Mitfords to Hitler." Diana admitted that having often met "drawing room communists," Hanfstaengl "was my first drawing-room Nazi."

Hanfstaengl glided through the British beau monde. He was a borderless socialite who claimed powerful connections to America as he was half American and where he was a Harvard classmate of Theodore Roosevelt Jr., the eldest son of Teddy Roosevelt, and became a regular guest at the White House. Putzi was renowned for playing the White House piano with so much zeal that he broke several strings. He was also distantly related to the German Saxe-Coburg family. All of this gave him the social credentials to represent Hitler as he traveled in august social circles in all three countries. George Messersmith, chief of the American Consulate in Germany during the early 1930s and later ambassador to Austria, remembered Putzi as "an excellent performer on the piano. Apparently, this brought him close to Hitler, who enjoyed certain types of music and liked to hear someone improvise on the piano. Because he had lived in the United States and claimed to have many friends there, as well as a deep knowledge of the country, Hitler made him one of his intimates and for several years he was one of the court favorites. Messersmith concluded that "Hanfstaengl has no prestige in the Party, but does have access to Hitler and is one of the few people who see him during his leisure moments."

Hanfstaengl's young son Egon was Hitler's godchild. In 2005, Egon described to the *Sunday Telegraph* his delight in playing with "Uncle Dolf": "I loved him. He was the most imaginative playmate a child could wish for. My favourite game with him was trains. He would go on his hands and knees, and pretend to be a tunnel or a viaduct. I was the steam engine going on the track underneath him. He would then do all the noises of the steam train."

In 1933 Hitler appointed Hanfstaengl *Auslandspresschef*, or foreign press chief of the National Socialist Party, a title that did not entail much formal responsibility except to keep him in the forefront of the type of social events where he would eventually meet Unity and Diana. Putzi was a controversial character, either loved or actively disliked. Messersmith went on to report that at one dinner party at the home of the American correspondent for the United Press, Putzi arrived late and groped the hostess's young niece under the table. Flustered, the niece eventually excused herself from the table and left the dinner in a flurry. Messersmith commented that Putzi "thought his position as Hitler's favorite and his uniform permitted him to do anything."

Later in the same report, Messersmith recounts a similar situation where Hanfstaengl was tormenting the young daughters of the Jewish hosts of a country weekend. Again, Messersmith stepped in to admonish him: "I said that he thought he could get away with anything but this was something he could not get away with. I told him that if he did not leave the daughters of the house alone, with whom he was taking these liberties, only because he thought he could get away with it, because they were Jews, that I knew what to do about it and I would certainly do it that same evening. Hanfstaengl knew what I meant and that he knew that I was going to tell Goering about it, who would not stand for that sort of thing." In the case of both Goering and Hanfstaengl, the trajectory of their careers would go in very different directions, at the führer's discretion.

Despite Messersmith's scathing observations, Hanfstaengl glided through high society in all the countries where he claimed an affiliation. Unity and Diana took Putzi up on his invitation and happily attended the 1933 *Parteig*, or Nazi rally, that began on August 1, 1933, in Nuremberg, as delegates of the BUF. Nuremberg was the site of many of Hitler's early speeches. The *Parteig* was the first in a series of yearly Nazi propaganda events, ending in 1938, to be held after Hitler's ascension to chancellor, and the fanfare of this first rally formally ushered in the Nazi era. Diana wrote home that Putzi "met us at the station (in Nuremberg). The *Parteig* turned out to last four days, not one as we had imagined. The old town was a fantastic sight. Hundreds of thousands of men in party uniforms thronged the streets and there were flags in all the windows." Unity added that she was looking forward to attending a performance of Wagner's *Der Meistersinger* as well as a fireworks display. Although Diana did not understand Hitler's speech, she reported that an "electric shock" ran through the crowd when he spoke.

While later annual *Parteig* rallies were used to cultivate and indoctrinate new supporters of the Third Reich, those invited to this first rally qualified to attend as a reward for their party loyalty. Over four hundred thousand party members plus one thousand guests traveled to Nuremberg on special trains outfitted in Nazi regalia for the party rally. A brochure published after the rally attested that "even the foreign guests are swept off their feet by the spirit and determination of the Hitler Youth."

Hitler commissioned Leni Riefenstahl to make the first in a series of propaganda films for the rally, *Victory of the Faith*, starring Hitler. The film was lost for many years and only recently rediscovered. The *London Observer* reviewed the film at the time, commenting, "The film is one long apotheosis of the Caesar spirit in which Herr Hitler plays the role of Caesar while the troops play the role of the slaves. It is certainly to be hoped that this film will be shown in all cinemas outside Germany, if one wishes to understand the intoxicating spirit which is moving Germany these days."

Unity gushed that the 1933 rally was "the most fascinating thing I have ever been to in my life." A photograph shows that Unity and Diana were the only women among the BUF delegation. Fellow delegates included William Joyce (later known as Lord Haw-Haw), one of the preeminent supporters of the BUF and author of *Dämmerung über England* (*Twilight over England*). In his memoir, Joyce wrote, "In 1933 I joined Sir Oswald Mosley's new movement, the BUF. I became one of the leading political speakers and writers of that movement; for three years I was Mosley's propaganda chief. These were marvelous times and I shall never forget them. I used all my influence in the movement to give the party a strongly anti-Semitic direction—and I may say that I succeeded in that direction." Other members of the delegation included Alexander Thomson, who was remembered by Mosley as an "exceptional thinker" and "one of the finest fighters for our cause we ever knew," as well as a decorated war hero, Captain Vincent.

Hanfstaengl noted in his memoir that Unity and Diana arrived in Munich during the summer of 1933 with a letter of recommendation from Otto von Bismarck. They left their calling cards at the Nazi Party Headquarters and Hanfstaengl promptly called them and invited them to attend the Nuremberg rally, an event that had been expressly prohibited by their parents. Upon her return to England, frustrated in her attempt to meet the führer, Unity admitted that "my mother was so furious that I had gone to the *Reichsparteig* that she said she would never let me go abroad again."

It was at the 1933 Nuremberg rally that Unity saw Hitler for the first time. "He looked so touching," she wrote to her family, that "the first time I saw him I knew there was no one I would rather meet." Unity was apparently not alone in her fascination. Women across Germany were enthralled by Hitler. He received hundreds of love letters each day at Nazi Headquarters in Berlin.

Although Putzi was able to produce tickets for Unity and Diana to attend the first Nazi Nuremberg rally, and introduce them to other high-ranking Nazi officials, he was not able to fulfill his promise to meet Hitler. For that, they were on their own. Unity and Diana got as close as the anteroom of the Deutscher Hof hotel where Hitler was staying. En route to the hotel, Putzi became increasingly aware of comments directed at the two sisters about their makeup, especially their red lipstick—contradicting the recent Nazi proclamation of German womanhood: blond and cosmetics-free. Eventually, Putzi ducked behind a building with them and produced his handkerchief, pronouncing, "My dears, it is no good—to stand any hope of meeting him (Hitler) you will have to wipe some of that stuff off your faces." To which Unity lamented, "I can't possibly do without lipstick."

The sisters returned to England without meeting Hitler, which only deepened Unity's devotion and determination. She vowed to return to Germany. Like many English debutantes of her vintage, nineteen-year-old Unity persuaded her parents to send her abroad for finishing school. In Unity's case, the most economical and best ideological fit was determined to be Madame LaRoche's finishing school in Munich where Unity arrived during the spring of 1934.

Germany was a logical destination for young British girls as many English aristocrats were of German descent and still had extended family in the country. Baroness Madame LaRoche's school, located in her large home at 121 Königinstrasse, was a popular destination for these young English debutantes, promising instruction in German language and the arts. There were also picnics and outings with suitable and eligible young German army officers. Unity decided that a German finishing school would be the most expeditious path to pursuing her infatuation with all things German, and most especially the chancellor. She also argued that a German finishing school had the added advantage of being less expensive than one in Paris, an attractive option to her parents.

One student at the LaRoche school remembered her time there fondly: "Munich was a new kind of life. . . . It was such a leisurely pace, with no pressures, no social whirl. People there went to the opera not because they wanted to be seen in a new evening gown, but because they loved the music. I was enchanted by the school and the parks and the modest, but dignified baroness."

Armida Macindoe, another fellow student, recalled that "the Baroness was easy-going . . . didn't appear at breakfast. Then Fraulein Baum would come, the governess who lived out, and we went with her to town. The Baroness taught nothing. Fraulein Baum taught German, we had piano, singing, painting, in afternoon classes to which we also bicycled. Lunch and tea were with the Baroness, supper in the schoolroom. The Baroness was asleep by nine o'clock and knew nothing if we went out. Annie, the maid, made the beds, cooked, and waited at table. Frau Hacha, the cook, was already old then."

Unity recalled that the school was "fascinating" with charming students and good food. She also had a suitor, or, as her sister Deborah referred to it, a "semi-romance" with Putzi Hanfstaengl. As Unity was older than the other girls, she had fewer academic requirements, which suited her well. Macindoe remembered admiring Unity, who arrived at the school six months after her. "I looked on her with awe, but not because I knew anything of her ideas. . . . I don't remember her having lessons like we did, she wasn't in the schoolroom. She was free-lance, she floated around. If she wanted something from the schoolroom she'd just come in and fetch it. She had her black shirt and it was taken in its stride. One was too polite to pass remarks. She was fully-fledged pro-Hitler, that's why she'd bothered to come out."

One impressionable debutante, Joan Tonge, wrote to her parents about the appeal of the German soldiers to the young English wards of Madame LaRoche. The German officers were "madly elegant, arrogant and conceited, and had tremendous presence. Their uniforms were immaculate and their self-esteem was Perspex strong." Another student at the school, Ariel Tennant, remembers walking through the Englischer Garten with her cousin, Derek Hill, a young British painter studying in Munich, and Unity. Unity grabbed Ariel by the arm and threatened to "give (it) another twist" if she did not admit to liking Hitler.

When Unity was not listening with joy to the Nazis singing outside her window as they walked by, browsing shops or reading in the Englischer Garten, she devoted herself to one single academic pursuit: learning German with Fraulein Baum. Her sister Diana claimed that "never before . . . had (Baroness LaRoche) known a girl like Unity, who set herself with a passionate single-mindedness to learn German so that when she met the Fuhrer, as she felt convinced she would one day, she would be able to understand what he said." Unity wrote breathlessly in a letter home that Putzi promised "on his hand of honor, no kidding," that if she learned to speak German, he would introduce her to Hitler.

Although he had recently been named chancellor, Hitler was quite visible in Munich in 1933. One of his favorite haunts, the Carlton Tea Room, anticipated his arrival with a permanent card on a prime table, RESERVIERT FUR DEN FÜHRER. Since Putzi had failed to fulfill his promise to introduce Unity and Diana to Hitler during the *Parteig* rally, Unity took matters into her own hands. Derek Hill knew about Unity's vow. In a letter to Diana, Unity wrote that he "rang (me) up from the Carlton Teeraum & said that

He was there. Derek was having tea with his mother & they were sitting just opposite Him. Of course I jumped straight into a taxi, in which in my excitement I left my camera which I was going to take to the shop. I went & sat down with them, & there was the Fuhrer opposite. The aunt said 'You're trembling all over with excitement' and sure enough I was, so much that Derek had to drink my chocolate for me because I couldn't hold the cup. He sat there for 1½ hours. It was all so thrilling I can still hardly believe it. If *only* Putzi had been there! When he went he gave me a special salute all to myself."

In June 1934, while pleasantries abounded for the young women at Baroness LaRoche's finishing school, the political horizon in Germany was darkening. After assuming the chancellorship the year before, Hitler further consolidated his power by executing those around him whom he deemed threatening. In one instance, he fabricated a supposed coup by Ernst Röhm, leader of the SA, the original paramilitary group of almost three million members that formed the Nazi Party. Röhm, who was dragged out of his bed along with other supposed political opponents, was either shot on the spot or soon after during what became known as the Night of the Long Knives. Unity, thrilled by the Night of the Long Knives, wrote Diana: "there were SS men dashing about the whole time on motorbikes & cars. It was all very exciting. . . . I am so terribly sorry for the Fuhrer—you know Rohm was his oldest comrade & friend, the only one that called him '*du*' in public. How one could do what Rohm did I don't know. It must have been dreadful for Hitler when he arrested Rohm himself & tore off his decorations. Then he went to arrest Heines & found him in bed with a boy. Did that get into the English papers? Poor Hitler. The whole thing is so dreadful."

Having escaped the watchful eye of both her parents and London society, and following a relatively unsuccessful debut, Unity enjoyed the freedom at Baroness Laroche's home. She also developed a friendship with Putzi's older sister, Erna, who referred to herself in the third person as "Miss Hanfstaengl." It was Erna who introduced Unity to all of the right Nazis and the cream of Munich society. This included Erna's cousin, Eberhard Hanfstaengl, the director general of a prominent Munich art gallery; Arno Rechberg, heir to a chemical manufacturing fortune; and the Baroness Redwitz, an heiress to the Bruckmann publishing family, publisher of the writings of Houston Stewart Chamberlain, and an early advocate of Aryan racial superiority. Unity was also introduced to Hugo Bruckmann's wife, Elsa, formerly Princess Cantacuzene of Romania, a devoted Nazi, who established the "Salon Bruckmann," where she introduced Hitler to prominent German industrialists.

While Unity was delighted to meet Erna's influential friends, Erna was also proud of her association with Unity. In fact, Unity eventually lived with Erna after giving up the pretense of finishing school and launching into her true mission for being in Munich, proximity to the führer. In a letter to Diana, Unity described Erna as "terribly sweet" and "enormously fat." Erna, in turn, wrote fondly of Unity: "On and off Unity lived with me practically the whole Nazi time. Her clothes were left in her room at Solln, a well-to-do area south of Munich, with her books and photographs. In the winter she would come to Utting (a fashionable town on the shores of Lake Staffelsee in the Bavarian Alps). Every summer, at least until 1939, she was my guest."

Unity was not the only family member drawn to life in Germany. Her brother, Tom, was studying in Berlin, spending weekends with Count Almásy at the Schloss Bernstein, possibly as a paying guest or even as Almásy's lover. Unity's mother, the Honorable Sydney Redesdale, visited Unity with her daughter Deborah and a cousin, Ann Farrer, also known as "Idden." She wrote that when she "went to Germany in July 1934 to visit Unity (she) was hoping to see Hitler." Munich was a vacation destination for Nazi Party members, a city known for carnivals and Oktoberfest, which was now decorated with swastikas. Unity regretted that "there may be no Nazi meetings to take you to, as all the party officials and S.A. and S.S. men are having holidays during July. It is a great pity as I would have liked you to have seen one of the big meetings, they are thrilling."

Fraulein Baum, Unity's German-language teacher, or "Baumchen," as Unity nicknamed her, was a devoted follower of Hitler and tipped Unity off that Hitler lunched most days when he was in Munich at the Osteria Bavaria on the Schellingstrasse. In contrast to the pomp and pageantry of Nazi rallies, Hitler's favorite Munich restaurant was quite modest. As Unity's friend Mary St. Clair–Erskine recounted, "we always went there for lunch, it was cheap. Herr Deutelmoser, the owner, we called Domodossola, a sweet gentle old bachelor. Bobo (Unity), like everybody, loved him. The two waitresses, Fraulein Rosa and Fraulein Ella, had been there for years and years. My husband's parents belonged to a club of painters and artists, the 'Klub der Lebenskunstler,' whose headquarters were there, so they had their Stammitsch and came every day."

Fraulein Baum invited Unity to join her there for lunch to see for herself. Unity wrote breathlessly that "suddenly everyone jumped to their feet and saluted and there he was in his sweet mackintosh. It was one of the most thrilling moments of my life." After that encounter, Unity made it her business to lunch most days at the Osteria, telling her mother that she drank only a glass of milk at other meals so that she could live within her budget, in hopes of meeting Hitler. She befriended both waitresses, Elsa and Rosa, who told her that Hitler mostly ate "vegetable cutlets."

Despite Fraulein Baum's virulent anti-Semitism, devotion to the Nazi cause, and tip about lunches at the Osteria Bavaria, Unity discovered that her teacher was half Jewish and then likely denounced her. "The most amazing piece of news of all is—Baum is out of the Partei! She was in the Osteria yesterday & Rosa told me. According to Stadelmann she was discovered to be a half-Judin (Jewess). Isn't it amazing. She also hasn't any work poor thing, as there was a big row in her Mutterheim at Starnberg & she was kicked out. I am really sorry for her, as the Partei & her hate for the Jews were really all she had."

Unity's school friend Armida Macindoe recalled going on stakeouts to the Osteria with Unity: "She used to go to the Osteria Bavaria restaurant and sit waiting for Hitler. She'd sit there all day long with her book and read. She'd say, I don't want to make a fool of myself being alone there, and so she'd ask me to go along to keep her company, to have lunch or a coffee. Often Hitler was there. People came and went. She would place herself so that he invariably had to walk by her, she was drawing attention to herself, not obnoxiously but enough to make one slightly embarrassed. But the whole point was to attract his attention. She'd talk more loudly or drop a book. And it paid off."

There were already stirrings of both surprise and frustration among Hitler's senior staff that Unity seemed to turn up wherever Hitler was in Munich, and most especially at the Osteria Bavaria. When her sister Diana Mitford came for a visit in 1934, Unity persuaded her to join in the pursuit of Hitler: "She (Unity) followed his doings in the newspapers, chatted to the doorman at the Brown House, looked to see if there was a policeman in the Prinzregentenplatz where he had his flat. If she considered it possible that he would turn up at the Osteria we lunched there. . . . Nothing would induce Unity to leave until he did. She willingly waited an hour and a half if necessary for the pleasure of seeing him go by her table on his way out." Even Unity's father, Lord Redesdale, a member of the House of Lords, waited with Unity at lunch at the Osteria when he visited during the late fall of 1934.

After months of stakeouts, on Saturday, February 9, 1935, Hitler finally took notice. He asked the Osteria owner, Herr Deutelmoser, to ask her to join him at his table. The twenty-year-old English debutante had found a place at Hitler's table, wearing full makeup, despite the official Nazi edict. Dietrich, Hitler's press chief, noted that "Hitler sent for Unity . . . and she had lunch at his table—thrilled to death of course." This incident also found its way to Dietrich's memoir, *The Hitler I Knew*, noting that "Hitler made the acquaintance of the English woman Unity Mitford . . . an enthusiastic follower of the British fascist leader Sir Oswald Mosley and a fervent admirer of Hitler. She had many private conversations about Anglo-German relations with Hitler, whose secret itineraries she usually guessed with great acuteness. Over the years Hitler frequently included her among the guests who accompanied him on his travels. She introduced Hitler to her father and her brother, when the two were passing through Munich."

Unity recounted the details of her first real meeting to her mother. "Hitler sat at one end (of the table) and at the other sat Bruckner, his adjutant. He had three adjutants, the second one was Schaub, who belonged to the working class and whom I liked very much. Dietrich, the Press chief, also sat at this table. He was very small and he told wonderful jokes in a very silent voice. Bormann was there too and he was an awful man, I did not like him."

Julius Schaub, one of Hitler's chief aides, later entrusted to destroy all of Hitler's personal letters before his suicide in 1945, told the German newspaper *Die Revue* in 1950 that while he was never sure why Hitler invited Unity to his table that day, he suspected it was because she so strongly resembled the ideal Aryan woman. Sefton Delmer, the German correspondent for the right-wing *Sunday Express*, and also a great friend of Unity's, described Unity as probably the only foreign woman in Germany to enjoy Hitler's acquaintance. Hitler referred to her as *"mein Sonnenschein,"* "my little sunshine." "Twenty years old, pretty with shining blue eyes and flaxen hair, she seemed . . . to embody the Hitler ideal of a Nordic woman." According to Schaub, Hitler was surprised to find out that Unity was English and not German. Schaub also noted that "with all her admiration for Hitler, the young Lady Mitford is quite clearly of the opinion that she is more or less his equal, although like everyone else she addresses him as *Mein Fuhrer*." Schaub also conceded that Hitler "was excited by the possibility of a love affair with her." Frau Salavatori, the wife of the owner of the Osteria, recalled that Unity "always parked that little car of hers illegally outside in the Ramburgstrasse, but the police knew it and never did a thing."

Unity described Hitler as wearing his khaki army uniform, swastika armband, Nazi Party badge, iron cross, a "hat he never puts on," and "a mackintosh, very old and grubby." She later wrote that at their first meeting they discussed his favorite film, *Caval-cade*, and Jewish interference in the Aryan state. Schaub added that Hitler looked at the copy of *Vogue* magazine Unity was carrying, along with a small German dictionary—and told Unity that England should never again have to enter into war on account of the "international Jews." They also discussed London architecture, film, and the new National Socialist architecture at Nuremberg.

After lunch, Hitler asked Unity to write her name on a scrap of paper for him and autographed a photo. Unity wrote to her father that "when one sits beside him it's like sitting beside the sun, he gives out rays or something. . . . If only you and Muv (her mother) could have heard him talk of world politics, of the League of Nations, of his own experiences. It was like a most beautiful speech, and yet he always wanted to hear what I had to say. He explains everything so clearly, and he speaks a lot in analogies." After her first lunch invitation, Unity said that having met Hitler, she was so happy that she wouldn't mind dying. Thus began a multiyear relationship between Unity and "Sweet Uncle Wolf."

Albert Speer also remembered meeting Unity at the Osteria Bavaria. "She was very romantic. The Osteria was a small inn, it is still there, and hasn't changed much. Small tables. There was a wooden partition and behind it a table to seat eight. An adjutant would phone the owner to warn that Hitler might be coming and to have the table clear. There was also a courtyard, with one table under a pergola and this was Hitler's favourite seat when the weather was not cold. Unity was quite often there, I was invited only every second or third time. Like me, Mitford would be invited by the adjutant Schaub. She was highly in love with Hitler, we could see it easily, her face brightened up, her eyes gleaming, staring at Hitler. Hero-worship."

Having finally succeeded in gaining admission to Hitler's intimate coterie at the Osteria Bavaria, it was not long before Unity's visiting family was included. In April 1935, Unity wrote to Diana that she "went to the Osteria & the Fuhrer was there. He sent Bruckner to invite me to his table, & I went & sat next to him. . . . The Fuhrer was so sweet & stayed a long time & talked a lot about all these Notes. He said he would like to see their mother. The next day (yesterday) Bruckner came to the Osteria to invite us to tea with the Fuhrer at the Carlton at 6. We went & there he was, and he said I must be the interpreter, but as you can imagine it was

very embarrassing as no-one could think of anything to say. . . .
Muv tactfully went away after about an hour, I stayed on & after
that of course all went swimmingly, he stayed until ¼ to nine. Of
course it was bound to be embarrassing with Muv as she can't
speak German, that is always rather a wet blanket. Whenever I
translated anything for either of them it always sounded stupid
and translated. On Tuesday by the way he asked after you, & sent
you Grusses (greetings). . . . I fear the whole thing was wasted on
Muv, she is just the same about him as before. Having so little
feeling she doesn't feel his goodness & wonderfulness radiating
out like we do, & even Farve (her father) did. She still says things
like 'Well I'm sure he is very good for Germany, but' . . . She will
admit that he has a very nice face."

Unity also hoped to introduce her father to Hitler, convinced
that Lord Redesdale sympathized with the führer's Fascist ideology. Unity's sister Jessica, a dedicated Communist, cynically
confirmed her parents' Nazi inclinations: "before long they would
prevail and Muv and Farve would be given a royal time in Germany. They would be lent a chauffeur-driven Mercedes-Benz,
shown all the gaudy trappings of the new regime and they return
full of praise for what they had seen."

Unity also made sure that when her only brother, Tom, visited, he was equally won over by Hitler, although he seemed to
object to the anti-Semitic aspect of the early Nazi propaganda.
Unity took Tom to the Osteria to meet Hitler: "we lunched with
the Fuhrer twice—Saturday & yesterday—and although I didn't
want him to meet him, I am quite pleased now. He adored the

Fuhrer—he almost got into a frenzy like us sometimes, although I expect he will have cooled down by the time he gets home—and I am sure the Fuhrer liked him, & found him intelligent to talk to." Tom was so taken by Hitler that he posed in a rakish photo, tucked away in Unity's correspondence, with the car that Hitler lent them for the week he visited.

By the spring of 1935, Unity's access to the führer was becoming well known, especially among the police detail that guarded him. Hitler's erstwhile mistress, Eva Braun, tucked out of sight at the Berchtesgaden, Hitler's mountain retreat in the Bavarian Alps, also became aware of Unity's constant presence and frequent late nights with the führer. Similar rumors of a new mistress had circulated a few years before about Hitler and Winifred Wagner, the English-born wife of Siegfried Wagner, the son of Hitler's favorite composer Richard Wagner. Winifred ran the Bayreuth Festival, an annual summer celebration of Wagner's music, from her husband's death in 1930 until the end of the war. Unity's importance in Hitler's life, though, seemed to be different. As her sister Diana noted, "most of the women he met were desperately shy and over-awed in his company. There were exceptions like Frau Wagner and Magda Goebbels, but anyone meeting him for the first time after he became the Fuhrer of Germany hardly spoke in his presence except to say '*Ja, mein Fuhrer*' or '*Eben, mein Fuhrer*' or '*Selbstverstandlich, mein Fuhrer*,' as the case might be. Unity was never awed in her entire life. She said what came into her head."

In an article in the *Sunday Dispatch* in January 1942, the American news reporter Ernest Pope looked back on the prewar years in Munich and his meetings with the young British aristocrat who had captured Hitler's attention. Pope described "looking up" Unity whenever he returned to Munich. She would receive him with a large black Great Dane by her side "who was as suspicious of me as was Unity. . . . Hitler's girl admirer has lovely hair, a beautiful complexion and a tall erect, full figure. But two features mar her Nordic beauty. Her eyes give an impression of chronic severity amounting to almost crossness." He continued that "Unity has spent everything from an afternoon to several days at Berchtesgaden. I was shown a photograph of her with Adolf in an informal tete a tete having tea in the Fuhrer's home."

According to Pope, Unity played a pivotal role in guiding Hitler's foreign policy with England, writing that until the Czech crisis of 1938, Hitler actively advocated for an Anglo-German alliance that gave Britain control of the sea while Germany was given control of the Continent. Pope concluded that "had such a European policy been embraced by Britain and Germany, it is quite possible that Hitler might have offered to embrace Unity before the altar and effect an alliance between this member of the English nobility and the lord of that National Socialist Party."

Albert Speer believed that Unity's access and proximity to Hitler was a source of conflict among his close staff. "For those close to Hitler it was a nuisance. . . . It was amazing that someone not German was around Hitler and could listen to details of party politics and far-ranging policy. Hitler made no secret of his thoughts and astonishingly a Britisher was sitting there and listening. . . . The others round Hitler were cautious and did not want to ask anything, but she was straight and said things Hitler didn't like. She had cheek. Hitler's line was to get along with the British. She pressed the point. He was sympathetic. They would argue and he appreciated frankness in her." Speer continued, "I never heard she was a spy. Somebody so close to Hitler must have been checked. Hitler's outspokenness was calculated, talking secrets knowing that rumours would spread." One member of Hitler's close circle explained that "Hitler used to say that Unity talks so much that whenever I have anything to announce to the world, I have only to tell her."

Winston Churchill, a distant cousin by marriage (Churchill's wife, Clementine, was a cousin of Lord Redesdale), invited Diana (and not Unity) to lunch to find out more about Hitler. She told Churchill that Hitler virtually controlled Germany and that he was "the person that everyone was interested in." It is baffling that British intelligence, well aware of Unity's proximity to Hitler, the only British citizen with access to the führer, never attempted to use her to obtain information on their behalf. Even the Nazis were skeptical.

Erna Hanfstaengl confirmed the distrust and often jealousy that Unity soon generated among Hitler's close Nazi circle. Erna claimed to "know how deeply hated she was by the party. . . . She was more prominent than some German leaders and they could not forgive that. Frau Himmler came into my shop and in the course of conversation she remarked on that 'good for nothing Unity.'"

While Unity's brother, Tom, might have been charmed by

Hitler, Unity's older sister, Nancy, a novelist, wanted nothing to do with Unity and Diana's right-wing politics, turning down an offer to stay with Himmler and visit Germany as his guest. "We were asked to stay with somebody called Himmler or something, tickets & everything paid for, but we can't go as we are going to Venice & the Adriatic for our hols. I suppose he read my book & longed for a good giggle with a witty authoress. Actually he wanted to show us a concentration camp, now why? So that I could write a funny book about them." Her latest novel, *Wigs on the Green*, published in 1935, satirized Mosley and the BUF and offered a thinly veiled skewer of Unity's preoccupation with Nazi Germany.

Unity not only dined with Hitler, but she was also his guest, discreetly, on overnight train journeys. She stayed with him in a sleeper on his special train; his habit of taking her on train journeys accounts for her nickname among his entourage of *"Mitfahrt"* (traveling companion). Unity also became a frequent houseguest of the Goebbels family, often for weeks at a time, where Hitler would make frequent and often unannounced visits. Mosley, who later married Diana Mitford at Goebbels's home, wrote that "Diana was very fond of Frau Goebbels, who, with her husband, was often at dinner with Hitler. Goebbels, one of Hitler's closest confidants and his Minister of Propaganda, had in private life an almost exaggerated sense of humour which, surprisingly, Hitler shared; it was one of the bonds between them." When defeat was imminent, Goebbels and his wife, Magda, poisoned their six children and took their own lives in Hitler's bunker.

Unity was also a frequent guest at Hitler's Munich residence.

As her school friend Mary St. Clair–Erskine recounted, "we went a lot to his flat in the Prinzregentenstrasse, it was round the corner. The curtains would be drawn and it was always dark. Lovely flowers everywhere. He ordered lots of cream cakes which he'd be disappointed if we left, but he'd eat a bit of *Knackbrot* (crisp bread). He had a great big globe on a stand; and liked to show us any new pictures he had. He was extremely nice and kind to us. A pleasant host."

Sir Oswald Mosley was also a lunch guest of Hitler's twice during this period, introduced by Joachim von Ribbentrop. As Mosley wrote in his memoir, "Hitler had solemnly introduced me to Unity Mitford at the luncheon he gave me in April 1935, as he was unaware that we knew each other." (Unity was the sister of his then mistress Diana so their paths had already crossed.) Hitler and Mosley met privately before the lunch and agreed that Britain and Germany should never go to war again. Mosley said that Hitler "was simple, and treated me throughout the occasion with a gentle, almost feminine charm. . . . He seemed to me a calm, cool customer, certainly ruthless, but in no way neurotic." Other guests at lunch included Winifred Wagner; Ribbentrop; Goebbels; former kaiser Wilhelm's daughter, Duchess Victoria Louise of Brunswick; and his granddaughter, Friederike, later the Queen of Greece.

As Unity's encounters with Hitler became more frequent, her embrace of the Nazi platform further solidified. Erna Hanfstaengl observed that "Unity took up anti-Semitism as breezily as everything else, talking about Jews with her usual exaggeration; she wanted to have them all burnt. 'Burn the Jews, that's the thing for them,' she would say, it was the fashion to chatter on like that."

Unity's increasingly public position in Hitler's entourage and extremist views began to be followed by the British broadsheets. The *Sunday Express* published an article about Unity on May 28, 1935, "She Adores Hitler: Daughter of a British Peer" with a large photo of Unity, noting her "Nordic" beauty. Unity gushed to the Special Correspondent that "the hours I have spent in his (Hitler's) company are some of the most impressive in my life. The entire German nation is lucky to have such a great personality at its head." Soon after, in a letter to her mother, Unity wrote that Hitler had told her that "Fascism must come in England, nothing can prevent it. The New Age has come." Unity saw herself as instrumental in laying the foundation between the two countries.

In Germany, Unity sat for a long interview entitled *"Eine Britische Fachistin Erzahlt"* or "Confessions of an English Fascist Girl" professing her Nazi beliefs, published in the weekend section of the prominent *Munchener Zeitung* later that June. The interview recounts how as a member of the BUF, Unity "distributes broadsheets in the London streets, and carries on propaganda for the fascist party wherever she can." Unity asserted that "our anti-Semitism has called Jewry in England to account." She ends the interview with the claim that "we British fascists have a lot to learn from Germany . . . the moment our Jewish enemies are ready to attack us—the time for this we know, is likely to be soon—then our struggle will at least reach its final decisive stage. The British fascist party is a party of front-line soldiers and youth. Hitler had his SA and Mosley has his 'Iron Front' and he himself once told me, 'I admire Hitler's national socialist movement: when at last we hold the helm of state in our hands, friendship between Germany and England will prevail."

The British ambassador to Germany flagged this interview in a reporting cable to the Foreign Office, using it as evidence of Unity's role in creating a link between the BUF and Nazi Germany. While it reveals that the Foreign Office had become aware of Unity's presence in Germany, her relationship with Hitler was noted as a curiosity. "You may like to know that the *Munchener Zeitung* of 22nd June contains an interview with the Hon Unity Mitford on the subject of Fascism in England. Miss Mitford, although only 20 years of age, appears to represent Sir Oswald Mosley at Munich and has been living there since September." It continues that "the Embassy does not seem to know about Miss Mitford's acquaintance with the Chancellor which is very curious indeed. She is a student in Munich—and it is true that her family are friends of Sir O. Mosley but I can't believe she 'represents' him. I understand that she sees Herr Hitler very often in Munich."

Unity took another intrepid step in support of the Nazi cause, speaking at a rally organized by Julius Streicher in Hesselberg in the hills outside of Nuremberg. Best known for carrying a whip, which he was known to use, Streicher was listed as Nazi #2 on his party card, compared to Hitler, who was Nazi #7. Streicher was also the virulently anti-Semitic founder of the Nazi newspaper *Die Stürmer*. It had a modest circulation of 28,000 when Streicher bought it, reaching a circulation of about 100,000 by mid-1935.

Streicher knew that Unity had Hitler's ear and was likely opportunistic in putting her on the stage where she was joined by other Nazi dignitaries including a deputy propaganda chief.

Unity described the scene behind the podium in a letter home: "It is the custom in Germany at midsummer midnight, to roll great burning wheels from the tops of the hills into the valley. It is a most extraordinary sight to see fiery wheels bowling along down the hill, getting faster and faster." The *Frankische Tageszeitung* reported that Streicher called Unity up to the swastika-festooned podium and presented her with a big bouquet of flowers, calling her "a brave English girl" and reading excerpts from the soon-to-be-published interview in *Die Stürmer*.

Unity made a few brief and impromptu remarks about her admiration of Streicher and her support of the Nazi cause, proudly proclaiming, "Everyone should know I am a Jew hater." The crowd of two hundred thousand Nazi Party supporters chanted *"Heil England"* as she left the stage. In a letter, Nancy Mitford commented that "we were all very interested to see that you were the Queen of the May this year at Hesselberg."

Unity's remarks at the Hesselberg conference garnered Hitler's attention. In a postcard kept by her mother and dated Hesselberg 1935, Hitler wrote: *"Du bist wunderschön. Du bist mein Ein und Alles. Mein Herz gehört Dir."* The note was translated in Unity's handwriting, "You are beautiful. You are my Everything. My heart will always belong to you."

Coverage in the conservative *Daily Telegraph* was far less complimentary, calling the Hesselberg gathering a pagan ritual and condemning the daughter of a member of the House of Lords for her comments. Unity responded to the criticism, pleading with them to "try to look at the whole thing from my point of view. I couldn't refuse Streicher's invitations could I, firstly for politeness sake, secondly because I naturally longed to go. When once there, I couldn't refuse the bouquet offered to me, nor could I refuse to go to the microphone when he called for me . . . Really I do think it is unkind to write like that, as if I'd done something wicked. Surely all that matters is that it shouldn't go against my own conscience, and it doesn't."

In Germany, Unity assumed the role of emissary between the English Fascist movement, headed by her soon-to-be brother-in-law Oswald Mosley, and the Nazi Party. In further interviews she gave to the German press she explained that "Oswald Mosley is our Fuhrer, which we English fascists are behind with the same enthusiasm the whole German people are behind their wonderful Fuhrer today. . . . We stand only at the beginning of a fight that the National Socialist movement in Germany has already completed victoriously." Unity also directly addressed the rising tide of anti-Semitism in England in a letter to the editor of *Die Stürmer*. "The general English public has no idea/knowledge of the Jewish danger. The English Jews are always described as 'decent.' Maybe the Jews in England are shrewder with their propaganda than in other countries. I do not know, but it is a well-determined fact that our fight is very hard. Our worst Jews work only behind the scenes. They never come into the public and because of that we cannot show them to the English people in their true terribleness.

We have an urgent need for newspapers like the *Sturmer* which tell people the truth. Hopefully, however, you will soon see that we in England will also become victorious against the world enemy, despite all its cunning. We look forward to the day on which we can say with force and authority: 'England for Englishmen! Out with the Jews!'"

While the British press was mixed in its response with the *Daily Mail* headline, "Peer's Daughter as Jew Hater," the reaction to her letter to *Die Stürmer* was widely praised in Germany. She wrote to her family that "I get a lot of letters of praise for my *Sturmer* letter every day, from all parts of Germany. Some of them send me little presents, and one young Austrian SS man who is having to return to Austria, where he will be put in prison probably for five years, came to see me and said he would like me to have his dagger and Swastika arm band. Poor thing. I was so sorry for him. The dagger is lovely."

The *Jewish Chronicle* was dismissive in its reaction to Unity's virulent *Die Stürmer* letter as a bid for attention by a young lady in need of "traditional discipline." It also accused the British press of publicizing her pronouncements because she is a member of the British aristocracy. In an editorial, the *Chronicle* observed that "she realized that without her title, which was, of course, a mere accident of birth, her vapourings would have been of no more concern to the public than those of any other irresponsible young hussy out to get her name in the papers."

However, the mainstream British press was more complimentary. The Germany correspondent for the *Gloucester Citizen*,

Michael "Mick" Burn, an old London friend of Unity's, wrote to his parents that Unity "is terribly grand and very nice. People think quite seriously that she is going to marry Hitler, who gives her lunch parties in his flat." Burn subsequently toured Dachau, an invitation that came in response to his book *I Was Hitler's Prisoner.* Burn commented that the prisoners of Dachau, often Jewish as early as 1935, represented more of a threat to Germany than the brutal SS soldiers who guarded them.

Burn was also one of many Britons who attended the 1935 Nuremberg Party rally as a guest of the German government. These annual gatherings of the Nazi Party in southern Bavaria gave Nuremberg the nickname "City of the Reich." The 1935 party rally was also where the Nuremberg Laws were announced for the first time. These laws effectively changed the legal status of German Jews from citizens to "Jews in Germany," establishing a framework for the persecution and, ultimately, the extermination of Jews not only in Germany but across Europe.

The spectacle of Nazi pageantry and propaganda at the annual *Parteig* conventions was electrifying. The *Frankische Tageszeitung* made special note of "Leni Riefenstahl and the two English Fascist girls who had come together on the podium (and) were greeted uproariously." Elmar Streicher, the son of the mastermind behind the *Parteig* spectacle, recalled that Unity had become part of Streicher's close circle as early as 1934. "Unity was always a lady, she kept her distance, but not with my father. At their first meeting, she had stood up in front of him like a general. She wore dark suits, cut like a uniform, usually her black shirt, and always the party badge. She also always wore big gauntlets. She was a bit of a suffragette."

By her own account, Unity's proximity and intimacy with Hitler continued to increase after the events of the summer. In a letter to Diana, Unity recounted, "The Fuhrer talked a good deal about the Jews and said that lately 10,000 have gone from Germany to Italy because there they can do good business in a way of profiteering. He is happy to be rid of them."

On Christmas Eve 1935, Unity returned home to her flat in Munich and "found in my room a lovely Christmas Tree, all decorated and with candles and standing in a huge basket quite full of boxes and chocolates, cakes, biscuits, fruit and nuts. On it was a card with an eagle and *Hakenkreuz* (swastika) printed on it, saying 'Frl Unity Mitford, Best Christmas and New Years wishes sends you faithfully Adolf Hitler.'" Hitler had a reputation for giving gifts to women. Among other gifts she received over the years included a camera, a small gold swastika pin, as well as various signed framed portraits of the führer. Unity's first biographer, David Pryce-Jones, claimed he was given these items as he was writing the book about Unity and left them with his father's papers. Alan Pryce-Jones, an intelligence officer during World War II who immigrated to the United States after the war, donated boxes of his personal papers and other memorabilia to the Beinecke Library at Yale University in 2012. Close recent inspection shows that Unity's items were not there.

In early January 1936, Unity wrote to Hitler's assistant, Gauleiter Wagner, for permission to visit Dachau. Wagner told her that women were generally forbidden to visit a concentration camp but made an exception for Unity, her sister Diana, and a British journalist friend, Mick Burn. The Dachau camp had opened outside of Munich in 1933 for "political" prisoners. While some visitors foresaw the horrors that were to come at Dachau as early as 1936, Unity and Diana did not register any discomfort.

The sudden death of King George V on January 20, 1936, and the ascension of his oldest son, Edward VIII, a rumored German sympathizer, provided an important bridge for the Nazi sympathizers in England. Unity made a point of visiting the office of the British Foreign Consul in Munich to sign a book of condolences, noting "a very sweet bit in a German paper" that she quotes: "Germany stands with the English people at the King's bier, and takes part in the grief which moves England today." She also attended a memorial service for the king on February 2, which she found disappointing. "The service was dreadful, because the parson was so awful & made a hopeless sermon in very bad English, mixing his metaphors so much that in the end one forgot whether the King was a temple or a pillar in the temple or a bird nesting in the roof of the temple."

CHAPTER 5

"SWEET DR. GOEBBELS"

Fascist Friendships

Right-wing movements that followed in the German Socialist example both proliferated and intensified in England during the decade of the 1930s. As relations warmed between English aristocrats and the top echelons of the Nazi Party, Jessica Mitford described the evolving attitude among English aristocrats as shifting from dismissive to mainstream. She originally saw her sister Unity's "newfound interest as rather a joke. Conservative opinion of Hitler at that time ranged from out-right disapproval of him as a dangerous, lower-class demagogue to a grudging sympathy for his aims and methods—after all, had he not decisively crushed the German Communist Party and destroyed the labor unions in a surprisingly short time?"

So many British citizens attended the 1935 Nuremberg *Parteig* that one Foreign Office official wrote in an intelligence brief that "there were a number of particularly silly people at Nuremberg." Among the "silly people" was Henry Williamson, a British naturalist of some renown and veteran of the Great War. Williamson believed that "Hitler was essentially a good man who wanted only to build a new and better Germany." Williamson returned from the 1935 rally especially impressed by the Hitler Youth, whose healthy pursuits he compared to the children in the slums of East London, from which many BUF members hailed. Like Mosley and Diana Mitford, Williamson was eventually accused of treason under Defence Regulation 18B, but unlike Mosley and Diana, who were imprisoned at Wormwood Scrubs although under favorable circumstances, Williamson was released after only two days. Williamson and Mosley remained lifelong friends.

Williamson's travel companion, John Heygate, whose reaction to the Nazi spectacle was more cautious than Williamson's, noted that Unity and Diana Mitford attended the rally as Hitler's personal guests, "arriving to see Hitler and his show, as if it might be the Lord Mayor's, or the Richmond Horse, or the Chelsea Flower Show, instead of a display of armed might designed primarily to rehabilitate Germany in despite of England."

Lady Irene Ravensdale, daughter of the Marquis of Curzon and sister of the late Cynthia "Cimmie" Mosley, attended the rally as Ribbentrop's guest. She was surprised by the effusive welcome she received. "Requesting Prince Bismarck at the German Embassy to secure me one ticket for one session, I was startled by being told

I would receive an official invitation from the Fuhrer himself for the whole session. This flung me into great agitation. . . . But when I discovered that many English people were going, already invited by von Ribbentrop . . . I accepted rather than be over-pernickety." Another attendee, Sir Arnold Wilson, observed that the event was "so simple, so solemn, so moving and so sincere as to merit, better than many customary religious rites, the title of worship. . . . If we cannot understand it and value it, we are the poorer thereby."

The propaganda campaign reinforced by parties and pageantry for the invited British guests garnered the desired response: good public relations for the Nazis in Britain. Former prime minister David Lloyd George held meetings with Hitler around the 1935 *Parteig*. He referred to Hitler as "the greatest German of the age" and expounded on "the advantage to Europe of strong men being in office." The positive impressions of the new Nazi regime spilled over into the British press. The *Sunday Chronicle* rhapsodized that "there is so much in the new Germany that is beautiful, so much that is fine and great. And all the time in this country we are being trained to believe that the Germans are a nation of wild beasts who vary their time between roasting Jews and teaching babies to present arms. It is simply not true."

While he did not attend the 1935 party rally, the Nazi debut on the world stage, Mosley maintained close contact with Streicher. Mosley sent a cable to *Die Stürmer* offices in Nuremberg, thanking Streicher for "your kind telegram which greeted my speech in Leicester. It was received while I was away from London. I value your advice greatly in the midst of our hard struggle.

The power of Jewish corruption must be destroyed in all countries before peace and justice can be successfully achieved in Europe. Our struggle to this end is hard, but our victory is certain."

By 1936, the Anglo-German Fellowship, formed the previous year to promote positive relations between England and Germany, with perceived sympathy for the Nazi Party, had over forty corporate members and a roster of England's most prominent citizens as members. Corporate members included Thomas Cook, Price Waterhouse, Unilever, and British Steel as well as members of the House of Lords, among others, the Duke of Wellington, Lord Londonderry, Lord Galloway, and Lord Nuffield, as well as Unity Mitford's father, Lord Redesdale. The society met at grand dinners where Ribbentrop, Hitler's newly appointed ambassador to London, convoked the Duke of Saxe-Coburg and Gotha, brother of Princess Alice and a frequent emissary for Hitler and Rudolf Hess, to mix with the top echelons of British society. Society heiresses such as Emerald Cunard, Lady Redesdale, Lady Sibyl Colefax, and Nancy Astor, doyenne of the stately home Cliveden, all vied for the opportunity to hold the next "at home." Both Guy Burgess and Kim Philby, later ringleaders of the Cambridge Five, infiltrated the Anglo-German Fellowship to furnish a firsthand account of the group's activities to MI5.

At its height in 1937, the Anglo-German Fellowship had approximately six hundred members. The leaders of both British society and government regularly organized shooting weekends at their country houses with visiting Nazis, discretely raising a swastika flag and drinking toasts to the führer. German guests of honor at these gatherings included Ribbentrop, Rudolf Hess, and the Prince of Saxe-Coburg and Gotha. There was a strong belief among those in attendance that the notables gathered under the mantle of the Anglo-German Fellowship could influence foreign policy toward Germany.

As Unity's father, Lord Redesdale, wrote in the *Anglo-German Review*, the magazine published by the society, in November 1936, "[H]as any one of (Hitler's) critics stopped to consider . . . what Europe would be like today if Germany had gone Red? By holding Bolshevism on the flanks of Western civilization, a tragedy was averted." This was just one headline attraction for their support of Nazi Germany. The *Review* also published enthusiastic pieces speculating about Hitler's love of England.

While the Anglo-German society thrived in England, its members also traveled to Germany. These trips were mostly organized by Ribbentrop, the recently appointed and very popular German ambassador to the Court of St. James's. Following his success in furthering German interests in England, Ribbentrop's next appointment was as Hitler's foreign minister for the duration of the war. Unity was skeptical about English aristocrats visiting Hitler, despite their pledges of support. In a letter to her sister Diana, she wrote that Hitler "told me that Lord & Lady Londonderry & the youngest daughter had visited him in Reichskanzlei last week. I felt bound to say that I was horrified that he should receive such people, and that he would soon find that practically all his English acquaintances were in concentration camps. He also admitted to having seen Beaverbrook, which horrified me even more. . . . After all, he isn't like any ordinary politician, who has to receive anyone who is important. Visitors to him should be reserved for those who have deserved it, by doing something for his cause or at any rate for really loving him, regardless of titles & money & importance, don't you think."

The Anglo-German Fellowship was also a pivotal force in convincing the British Football Association to host the German national team to a match versus England in December 1935. The game was played in a largely Jewish neighborhood at White Hart Lane, home to the Tottenham Hotspurs. Despite concerns about possible demonstrations and accompanying violence, a handy 3–0 victory by the British team softened the protests. Lord Temple, speaking at an induction dinner for the Anglo-German Fellowship, reminded those who had sought to cancel the game to "mind their own damn business. . . . The Germans have always been our good friends. They have always fought fair in war. . . . If another war comes . . . well, I hope the partners will be changed." At an elaborate celebration dinner after the match, Hans von Tschammer, the visiting Reich minister for sport, gave a speech saluting "the blue sky friendship between the two Nordic countries."

Similar invitations to the 1936 Olympics were issued to prominent Brits. The games, to be held in Berlin, provided another opportunity for the Nazi Party on the world stage promising to showcase the most lavish Olympics in the history of the games. On a stage adorned with the giant flags emblazoned with the black and red swastikas of the Nazi Party, Hitler opened the games with thousands of spectators and athletes in attendance. One British guest observed that "one was conscious of the effort the Germans were making to show the world the grandeur, the permanency and respectability of the new regime." The event began with a screening of the two-part, four-hour propaganda film *Olympia* by Leni Riefenstahl, a film with a generous budget celebrating the beauty of the human athletic form and its reflection in the ideals of the Nazi Party. After the war, Riefenstahl claimed that the film was financed by the International Olympic Committee, although payments have been traced directly to Goebbels.

Threats of a British boycott quickly receded as forty-nine countries agreed to participate. The event was broadcast in twenty-eight different languages from elaborately equipped transmitter vans located close to the Olympic stadium. The controversy surrounding attendance by the British team, which eventually sent forty-two athletes to the games, made it difficult to find last-minute funding. The team resorted to asking for donations from private individuals, like Lord Nuffield, and a ragtag compendium of British companies such as Imperial Tobacco and Horlicks Malted Milk as well as the Greyhound Racing Association. In the end, the British turned in a mediocre performance, winning fourteen medals.

The Games, proclaimed a great success for the international status of the Nazi Party, ended with the same fanfare with which it had started. A rousing rendition of Handel's Hallelujah Chorus accompanied the release of a thousand pigeons. The young Oxford don and later Labour MP Richard Crossman attended the games and noted that "Hitler, largely thanks to British foreign policy, has won the masses." Teams streamed by Hitler on the dais, some, according to Crossman, offering an "Olympic salute" or, like the French, more likely a "Heil Hitler." The *Daily Express* suggested that the British athletes were disrespectful toward their host by only nodding in Hitler's direction as they filed by. "It would not have done the British any harm if they had made a gesture for the country housing the games by following the unexpected example of France," pronounced the influential British daily.

Members of the Anglo-German Fellowship, among other right-wing British organizations, were delighted to receive coveted VIP invitations to the 1936 Berlin Olympics. As one British writer observed, "[A] section of smart London society was . . . not only pro-German but pro-Nazi." Ambassador von Ribbentrop sent hundreds of official invitations to the British society figures he had only recently met. And they were eager to attend. The American-born, British bon vivant, Sir Henry "Chips" Channon, and his wife, a member of the Guinness family, were among the first to accept the invitation. Channon was criticized by the prominent travel writer Robert Byron, who cynically noted, "I supposed I should not be surprised that you are prepared to sacrifice the interest of your adopted country in the supposed interests of your adopted class." Lord and Lady Brunswick were honored at a pre-Olympics dinner held at the Berlin chapter of the Anglo-German Fellowship. Among those in attendance were the Conservative politician Lord Zetland, secretary of state for India; Lord Mount Temple; Lord Rennell; and Lord Lothian, newspaper editor and former private secretary to Prime Minister David Lloyd George.

For the Olympics, Hitler was the houseguest of his minister of propaganda, Goebbels, "Sweet Dr. Goebbels" as Unity referred to him in a letter home, and his wife, Magda, along with her sister, Diana. The Goebbels home on Lake Wannsee had recently been appropriated from a Jewish family that had fled Germany. Already, the Nazi inner circle was suspicious of Unity's access to Hitler and the inner workings of the Nazi Party. Magda had been the one to invite the sisters to be their houseguests against Goebbels's wishes. He recorded the event in a diary entry. "I had a row with Magda about the visit," he wrote. "She cried and I was so sorry." Both Mitford sisters developed a close relationship with Magda, although their friendship never extended to her husband. Despite his "marvelous blue eyes" and "lovely speaking voice," Diana, in particular, was aware that Goebbels did not approve of their access to the führer. Possibly as a result, the sisters were lodged in a guest-house and not in the main house.

Chips Channon fondly recalled the pageantry of the Games, the chauffeured Mercedes limousines driven by storm troopers, whisking foreign guests to sports matches through a gray drizzle, and the elaborate parties that followed. Trees along the six-mile route from the Brandenburg Gate to the Olympic Stadium were decorated with Nazi flags: "Berlin has not known anything like this since the (First World) War. . . . It was fantastic, round-abouts, carts with beer and champagne, peasants 'dancing' and 'schuh-plattling,' vast women carrying pretzels and beer, a ship, a beer-house, crowds of gay laughing people, animals, a mixture of Luna Park and White Horse Inn. Thousands of pigeons were released when Germany won a medal."

Ribbentrop hosted a party for six hundred of his invited guests held at the Dalheim Villa. At that party, Channon described Ribbentrop as host to be "like the captain of someone's yacht, square, breezy, and with a sea-going look. . . . He is not quite without charm, but shakes hands in an over hearty way, and his accent is Long Island without a trace of Teutonic flavor."

Goering's party for two thousand that followed was even more elaborate. Docks were floated on the lake and guests were assisted into boats by beautiful young dancers carrying torches and ferried to a nearby island lit by lanterns, serenaded by orchestras, dance performances, and "fireworks on a scale which would have impressed the Romans."

For the Mitford sisters, and Unity in particular, Hitler's invitations to the most prestigious events of the summer of 1936—the Bayreuth Festival, the *Parteig* annual meeting of the Nazi Party, and the Olympic Games—cemented her in the führer's inner circle. In a letter Diana wrote to Unity after the *Parteig*, she gushed about "how sweet the Fuhrer was. He came into the room and made his beloved surprised face, and then he patted my hand. . . . Then I said we loved the wonderful parades and he said it was the best *Parteig* he had ever had because everything had geklappt (worked)." While Hitler stood for over five hours to review over one hundred thousand German soldiers who marched through the narrow streets of Munich in a show of Nazi power, the young British Unity Mitford stood in Hitler's shadow.

One of the most conspicuous examples of Hitler's increasing sway among influential British visitors was the marriage of Diana

Mitford to Oswald Mosley, hosted by the Goebbels. While the British gentry made the 1936 Nazi events the highlight of their social season, Oswald Mosley remained in England, organizing demonstrations that culminated in the infamous riots that took place in a Jewish neighborhood in the East End of London. At one, known as the Battle of Cable Street, anti-Fascist demonstrators clashed with members of the BUF. Led by Mosley, who was determined that "fascism can and will win Britain," and describing the Jewish enemy as "rats and vermin from the gutter of Whitechapel," the BUF took the Fascist cause to the doorstep of those they objected to most. A police force seven thousand strong met the Blackshirt marchers and culminated in the arrest of eighty-three protesters.

Shortly after the Cable Street debacle, on the evening of October 5, Mosley boarded a plane for Berlin where he was scheduled to marry Diana Mitford. Mosley had long feared that marrying Diana would detract attention from the BUF cause. A well-documented philanderer, Mosley had not been deterred by Diana's marriage to Bryan Guinness, regularly attending their parties on Cheyne Walk that were always a highlight of the social season. Notwithstanding her two young children and husband, Diana declared, "I'm in love with the Leader and I want to leave Bryan." Additionally, Mosley's Fascist politics and increasingly virulent anti-Semitic rhetoric had gained him many enemies that he wanted to protect Diana from. However, Mosley insisted that Diana must surrender her life to his Fascist cause, which he believed was the only way to save England. Diana's parents were not in favor of her divorce and especially her remarriage to Mosley, who they found to be unworthy of their daughter. They both agreed that a secret wedding outside the London spotlight would be best.

Diana's first thought was to get married in Germany. She learned from the British consulate that there was a reciprocal arrangement for weddings between Germany and England. Plans were quickly made for the ceremony to be hosted by the Mitfords' dear friend Magda Goebbels.

Unity and Magda Goebbels served as Diana's witnesses and the ceremony was held in "an ordinary middle-class drawing room, very bright and clean," as the bride noted, with a reception afterward at Goebbels's house on the lake. The wedding was attended by Mosley's best man, Robert Gordon-Canning, and one other friend, Bill Allen, as his witness. There was an aide bearing armfuls of roses and carnations, Goebbels's sister, and the guest of honor, Hitler. Diana wore a pale gold tunic over an ankle-length black skirt.

Diana wrote to Unity that the wedding was "the loveliest and at the same time the most terrible day for me. The wedding itself was so beautiful, and the blick (sight) out of Magda's window of the Fuhrer walking across the sunny garden from the Reichskanzlei was the happiest moment of my life. I felt everything was perfect. . . . The Fuhrer's orchids and Widemann's roses . . . and the ceremony, and the Fuhrer's wonderful present and the drive to the Schwanenwerder, and the wonderful essen (food) and Magda's and your sweetness too." Hitler gave the newlyweds a silver framed photograph of himself, embossed with the initials AH and the dual eagles of the German Reich. While the photograph and the frame were destroyed in a housefire in the 1950s, Diana cherished the wedding gift given to them by the Goebbelses, a pink leather-bound set of the full works of Goethe, inscribed by Magda:

> My dear Diana
>> In reminiscence of 6 October 1936
>> And the happiness of being with you both on your wedding day
>> With very loving wishes
>> And everything for your success and happiness
>>> Magda Goebbels

After the wedding lunch, Hitler gave a rousing speech in front of a crowd of twenty thousand at the nearby Sports Palace to inaugurate a National Socialist welfare plan. Unity said that the speech was "one of the best I ever heard him make." The *Sunday Express* reported that in this speech, Hitler launched an attack on Bolshevism and democracy, both of which he believed posed a threat to international peace and security. "Like Dr. Goebbels who spoke before him, he used bitter words about the foreign Press, and declared that 'these ridiculous newspapermen could squirt their poison.'"

Another wedding dinner was held that evening, hosted by Hitler himself. It was also the last time that Mosley saw Hitler. Two days later, Mosley returned to England, where he kept his marriage to Diana Mitford Guinness a secret. It remained so for two years, until November 1938, when both the *Daily Telegraph* and *News Chronicle* reported that the two had been married in Hitler's private offices in Munich. It wasn't until the birth of their first son that December that Mosley confirmed their marriage.

Diana wrote to Unity, who had left with Hitler the day after the wedding, that she was "sitting in a bower of orchids envying you, because I expect you are still in the Fuhrer's train." Unity responded that "he was sweet in the train last night & we had a lot of jokes."

At the wedding, Hitler and Mosley only exchanged platitudes with the help of an interpreter. Diana, however, managed to gossip with Magda Goebbels about the scandal of the day, Edward VIII's love affair with Wallis Simpson.

While history has trickled down that Edward abdicated because of his love affair with the twice-divorced American Wallis Simpson, there were perhaps more pressing political reasons for the prime minister to force Edward to renounce the throne, such as rumors of his intent to negotiate directly with Hitler. Edward told his cousin, the Duke of Saxe-Coburg and Gotha, that he viewed an Anglo-German alliance as "an urgent necessity" and insisted, "I myself will talk to Hitler, and I will do so here or in Germany. Tell him that please!" When the war ended, the Mosleys lived the rest of their lives outside of England, mostly in Paris, as neighbors and close friends of the Duke and Duchess of Windsor.

CHAPTER 6

WHOSE SIDE?

The Other Story of the
Duke and Duchess of Windsor

The Duke and Duchess of Windsor set sail on the SS *Berkshire* from Nassau in the Bahamas to Florida in April 1941. The former British king and his new wife took one of the last regularly scheduled trips between Nassau and the US mainland before the ship was requisitioned for US government use during the war. Ostensibly, the purpose of the trip was a dentist appointment for Wallis. What emerges from this six-day trip, meticulously surveilled mostly by J. Edgar Hoover at the request of President Franklin Roosevelt, reveals the concerns shared by both the Americans and the British about the duke and duchess's complicity with the Nazi high command.

A deputy to the duke in the Bahamas claimed that the duke was afraid to travel to Florida because he believed that he would be kidnapped by the Germans and traded for Rudolf Hess, a leading member of the Nazi Party, who had landed in Scotland in 1941 with the intent to make a deal with members of the British government who sympathized with the Nazi regime in order to avoid a war between the two countries.

The couple had arrived in the Bahamas on August 18, 1940, spirited away on orders from Churchill from Nazi-controlled Madrid and Lisbon where they were lingering after Wallis grew "bored" with the South of France. According to Nazi war plans, former ambassador to London and now foreign minister Joachim von Ribbentrop had been plotting to reinstall Edward on the British throne as a Nazi puppet, with Wallis as his queen. This end would be accomplished by capitalizing on the Windsors' established sympathies with both Hitler and the Nazi regime.

Hitler had long sought an alliance with the British monarchy through a union between the kings of England and the emperors of Germany. Earlier he had promoted the idea of a marriage between the Prince of Wales and Princess Friedericke, the seventeen-year-old daughter of Duke Ernst August of Brunswick and Duchess Victoria Louise. The duchess was horrified by the idea of her young daughter marrying Edward, as she herself had been rumored as a potential wife for him.

As a choice for a wife, Wallis Simpson was equally unappealing to the British royal family. Already twice married and American born, Wallis Simpson met Edward in 1932, when she and her husband, Ernest Simpson, were invited to join a shooting weekend at the Leicestershire home of Viscountess Thelma Furness and her husband. Simpson's financial pressures had already contributed to Wallis's roving eye. Soon after the weekend party, she invited the future king to dinner at the Marylebone flat on George Street, 5 Bryanston Court, that she shared with her husband. Wallis was a sought-after dinner party hostess, her table a destination where some of the most influential names in British society and right-wing politics, like Oswald Mosley, were frequent guests. That January evening, the prince stayed until 4:00 A.M. as the conversation drew him into a discussion of "new ideas that were bubbling up furiously in the world of Hitler, Mussolini, Stalin and the New Deal." As the duke would recall, "Wallis was extraordinarily well informed, her conversation deft and amusing."

By 1934, Wallis's husband was suffering ever greater business reversals, which accelerated her pursuit of the duke. She was sufficiently entrenched in Edward's circle that Thelma Furness, leaving for an extended trip overseas, met Wallis for tea at the Ritz Hotel before her departure and asked Wallis to look after "her little man" in her absence. Returning a few months later, Thelma wrote that indeed, Wallis had "looked after him exceedingly well." The resulting domestic situation, as Chips Channon, a diarist of the London beau monde, noted was the "ménage Simpson."

Wallis had a wandering eye and Edward was not the only opportunity she entertained. Joachim von Ribbentrop, who presented his credentials as German ambassador to the Court of St. James's in June 1936, also sparked her attention. Ribbentrop arrived in London with much fanfare, viewing his role as a public relations opportunity at the Foreign Office, whose legions were skeptical of Hitler. One fellow diplomat commented that Ribbentrop was a good choice for the job at the Court of St. James's, suggesting that he enjoyed the company of prominent people.

Ribbentrop's approval for the post of ambassador was accelerated through the usual vetting process, with relief that the German government had not only filled the vacancy, but with somebody well known and seemingly well-liked by them. Foreign Minister Anthony Eden had discussed the newly proposed German ambassador with the king. "Herr von Ribbentrop is so well known already it is perhaps unnecessary to give you the usual details about him," especially since the secretary of state was recommending that Ribbentrop be accepted that same day in order for an official announcement to be made before Eden went out of town at the end of the week. When Ribbentrop presented his credentials to King George VI in early February, his resonating greeting of "Heil Hitler" was criticized even by Goering, who later said that "because of that stupid tactlessness" he had risked becoming "persona non grata" with the British. Ribbentrop may have brought controversy with him to Britain but the British government was willing to overlook it. A brief memo written by Eden marked "Personal" to "Your Excellency" on the topic of Ribbentrop in November 1936 further revealed his determination to safeguard Ribbentrop's position in London. The memo suggested that no action be taken on an unnamed matter relating to Ribbentrop, "since such action must inevitably result in further publicity and controversy. I am sure this is the last thing which any of us desire."

In Edward, Ribbentrop had an easy mark. Over the years, the duke spoke fondly of his German ancestry, proclaiming to Diana Mitford that "every drop of blood in my veins is German." Diana Mitford Mosley recalled Edward mentioning that the royal family would wait until their guests left a gathering, at which point the family "comfortably lapsed into German." It was widely reported that the duke considered German to be his mother tongue and corresponded with his extended German family *"auf deutsch."* One cousin reported that after a trip to Germany in 1935, the future king took to "wearing a German helmet and goose stepping around the living room, for what reason I cannot imagine." In his defense, another friend remarked that since England had cut him off after he abdicated the throne, it was only natural that he would pivot toward Germany with which he had always felt a great affinity.

Hitler, in turn, saw an opportunity in Edward as heir to the throne. As early as 1936, Hitler was reported to be watching newsreels of the future king before the funeral of George V. To represent Germany, Hitler again sent Carl Eduard, Duke of Saxe and Coburg, a first cousin of the deceased king. This was one of ten trips that Saxe and Coburg made to England on Hitler's behalf during the period leading up to Britain's entry into the war.

The fact that Edward was forced to abdicate the throne in order to marry Wallis attracted the interest of the German regime. Rumors about his pro-German sentiments quickly circulated among the Nazi Party. The former Austrian ambassador to England, Count Albert von Mensdorff, reported that the duke was interested in returning to the throne if the Germans successfully invaded England, a development that he found "interesting and significant." As the duke planned his wedding to Wallis in Paris, in June 1937, the British ambassador, Eric Phipps, cabled the Foreign Office speculating about the duke's plans: the duke "seems to be rather embarrassed by the excessive attentions of the French, and compares their behavior unfavorably with the discretion with which he was treated in Austria." In his memoir, Albert Speer, the architect of the 1936 Olympic Stadium and other Fascist Nazi landmarks, wrote about the role the duke would have played in the Nazi regime: "I am certain that through him permanent friendly relations with England could have been achieved. If he had stayed, everything would have been different. His abdication was a severe loss for us."

Another British diplomat, Sir Robert Lockhart, noted in his diary after meeting Edward at St. James's Palace that "the Prince of Wales was quite pro-Hitler and said it was no business of ours to interfere in Germany's internal affairs, either re Jews or re anything else and added that dictators are very popular these days and we might want one in England before long." Ribbentrop also quickly recognized the advantages of the duke's pro-German position and was eager to exploit it: "I am convinced his (Edward's) friendly disposition towards Germany will have some influence on the formation of British foreign policy."

Ribbentrop's ambassadorship had all the hallmarks of great

success, a sworn allegiance and backing of Hitler, and a popular choice at the Court of St. James's. He also had the advantage of social charms, both cultivated and appreciated by the British aristocracy. As the German diplomat Paul Schwarz recounted, "Ribbentrop was a snob with a vengeance. To twist his stomach with caviar in the presence of the Duke of Devonshire or the American Ambassador he would walk more than a mile. . . . There is no genuine trait in him."

Ribbentrop saw an opportunity to infiltrate the heart of British tradition and political power through the Duke of Windsor. His first line of approach was through Wallis Simpson. Ribbentrop met Wallis through another American society hostess, Lady Emerald Cunard. He was a guest of honor at a dinner at her home on Grosvenor Square in June 1935. The two met privately again soon after. Ribbentrop sent Wallis seventeen carnations every day, delivered to her flat in Bryanston Court by a shop attendant. These carnations and her affair with Ribbentrop quickly became the main topic of conversation at the German embassy. Wallis was in "constant contact" with von Ribbentrop. In practice, "the Duchess was obtaining a variety of information concerning the British and French government official activities that she was passing on to the Germans."

As Ribbentrop infiltrated the Windsors' close circle, sensitive information about England began to leak in German diplomatic circles. One German diplomat pointed a finger not only at Edward, but at Wallis's access to her husband's papers. "Berlin was filled with loose talk about Edward. It was said that he neglected his duties in handling of official documents. Secret Ambassadorial reports were especially emphasized. At Fort Belvedere the Foreign Office dispatch bags were said to have been left open and it was possible that official secrets had leaked out. Was it not true that in Nazi Berlin one could hear stories which in London passed for State Secrets. There can be no doubt that some people in official British circles were aware that these rumors cast an undesirable reflection upon their King."

One of the duke and duchess's first trips as a married couple was to Germany. Sir W. Selby, British ambassador to Germany, wrote to Ambassador Eric Phipps in Paris where the Duke and Duchess had set up residence, that he had received "a letter dated September 20th from the Duke of Windsor informing me that he and the Duchess are visiting Germany in the near future 'in order to see what is being done to improve working and living conditions of laboring classes in several of the larger cities.'" Selby wanted Phipps to tell the duke that he would not be in Germany during the proposed dates, hoping that might head off an awkward visit by a former monarch with suspicious political allegiances.

Ambassador Phipps then updated Foreign Secretary Anthony Eden in a cypher telegram noting that "His Royal Highness told me he was leaving for Berlin on October 7th and expects to stay there for a couple of days. After that he will travel about in Germany, he does not quite know where, and will study housing and working conditions there. I warned his Royal Highness that the Germans were past masters in the art of propaganda and that they would be quick to turn anything he might say or do to suit their own purposes. He assured me he was well aware of this, that he would be very careful, and would not make any speeches."

News of the duke and duchess's trip also came through the "private channels" of the British embassy in Berlin. Cables exchanged between the various British consulates lamented the breach of diplomatic protocol in that the trip had been organized by the Reich command without consultation with the British embassy and confirming that "all costs of the trip would be paid by Germany."

The British government quickly shared news of the impending trip with its American counterparts likely because America was the duke and duchess's next destination after their trip to Germany. In a cable to Sir Alexander Hardinge sent to Balmoral Castle, Robert Vansittart, a former private secretary to the prime minister, and current undersecretary at the Foreign Office, insisted that although he was letting the prime minister know, "[I] imagine that his view, as in the case of the German visit, will be that nothing can be done to prevent it but that his Majesty's Representatives should not take any action which could be regarded as countenancing it." He added that such a tour, arranged without consultation with the Foreign Office, was "a bit too much" and that he hoped that "our missions abroad will be instructed to have as little as possible to do with them." Similar instructions were issued to embassy officials in America, although a discussion followed that it would be noted by the local press if consuls were not allowed to meet with the duke and duchess at their various American stops.

The response to the duke and duchess's proposed trip to Germany was much fretted over and is well documented in the Foreign Office classified archives. It was reported that, while the trip could not be prevented, care should be taken to stop H. M. representatives from taking any action which could be regarded as countenancing the visit." A recommendation was made by Phipps, the ambassador to France, that one way for the British government to distance themselves from the trip would be to have a member of the embassy staff meet the duke and duchess at the train station in Berlin. Even this proposal was met with indignation, cautioning that it would give the impression that the duke was traveling on official British business. Further instructions were given that members of embassy staff should not accompany the duke and duchess on any of their visits and that it would be "preferable" that they not even be entertained privately in the embassy, especially given the convenient absence of the ambassador. Both the ambassador and embassy staff were instructed not to accept any invitations by the duke or the German government during the visit, although if the duke invited the ambassador, he "should not decline." The embassy requested that, given the expectation that the visit would be front-page news and likely the subject of the prime minister's weekly public questions in Parliament, it should be clearly stated that the British government "in no way approves of the visit." The matter also escalated to the Prime Minister's Office, and noted Chamberlain's view "that it is not possible to stop the Duke of Windsor going to Germany."

The duke and duchess arrived by rail in Berlin on October 12 and were whisked off to lunch at Carinhall, the country estate of Hermann Goering and his second wife, a German actress, Emmy Sonnemann. While in Berlin, the Windsors also met SS Commander Heinrich Himmler, Rudolf and Ilsa Hess, as well as Joseph and Magda Goebbels, who remarked that the duke was "a charming, likeable chap; open, clear with a healthy common-sense approach, an awareness of contemporary life and social issues. . . . What a shame he is no longer King! With him an alliance would have been possible. . . . A great man!" On October 20, the duke's cousin, Carl Eduard, Duke of Saxe-Coburg and Gotha, hosted a dinner for the duke and duchess at the Grand Hotel in Nuremberg. Attended by the princes of Germany, dinner was an especially memorable event for the duchess. As she stood at the head of the receiving line, it was the first time that ladies curtsied and addressed her as "Your Royal Highness."

In addition to a full schedule of social engagements during their eleven-day trip, the Windsors also visited a concentration camp. Dudley Forwood, a member of the duke's equerry who traveled with the duke and duchess, noted that "we saw this enormous concrete building which I now know contained inmates: The Duke asked 'what is that' and our hosts replied, 'it is where they store the cold meat.'"

The highlight of the Windsor visit to Germany was lunch on October 22 at Berchtesgaden, Hitler's private residence in the Bavarian Alps. The duke and duchess traveled by private train, accompanied by Hitler's deputy, Rudolf Hess. They were met by a convertible Mercedes-Benz at the station and completed the journey up the mountain escorted by a motorcade of armed Nazi officers and police. Hitler met them dressed in his brown SS jacket. Wallis recalled that "his face had a pasty pallor and under his mustache his lips were fixed in a kind of mirthless grimace. Yet at close quarters he gave one the feeling of great inner force. His hands were long and slim, a musician's hands, and his eyes were truly extraordinary—intense, unblinking, magnetic, burning with the same peculiar fire." A translator, frequently corrected by the duke, was present throughout the meeting for the duchess's benefit.

According to the *New York Times* coverage of the visit, the duke gave a Nazi salute to Hitler upon arrival at Berchtesgaden. The same article recounts that Wallis was "visibly impressed with the Fuhrer's personality and he apparently indicated that they had become fast friends by giving her an affectionate farewell. Hitler took both their hands in his, saying a long goodbye, after which he stiffened to a rigid Nazi salute that the Duke returned." The duke admitted to the Nazi salute and defended it: "I did salute Hitler, but it was a soldier's salute." Regardless, Churchill congratulated the duke on a visit conducted with "distinction and success," despite the prohibition of any government involvement. The *New York Times* coverage of the German trip noted that Germany had lost "a firm friend, if not indeed a devoted admirer on the British throne." One outcome of the visit was the establishment of a line of secret communication between Hitler and Edward, passed back and forth by an intermediary. In these cables, Hitler addressed Edward as "EP," shorthand for "Edward Prince."

The British ambassador to Washington, Sir Ronald Lindsay, was less sanguine than Churchill about the trip, describing it to the US secretary of state as evidence of a "semi fascist comeback in England." At the request of J. Edgar Hoover, the FBI undertook an investigation of the duke. A September 1939 report to Hoover by Agent Edward Tamm concluded that "for some time the British Government has known that the Duchess of Windsor was exceedingly pro-German in her sympathies and connections and there is strong reason to believe this is the reason why she was considered so obnoxious to the British Government that they refused to permit Edward to marry her and maintain the throne."

The Windsors fled their home in Paris when France fell to the Nazis, leading them to Spain and eventually as houseguests of a pro-German Portuguese banker and Nazi agent in Lisbon during the summer of 1940. Following the fall of France in June 1940, Franco proposed to Hitler that Spain join forces with Germany. When the Windsors arrived in Madrid, the city was Nazi administered with an eye to enlisting Edward to assist the Nazi regime in what they believed would be the inevitable and swift conquering of Britain. According to documents declassified in 2003, Germany considered the duke to be "the only Englishman with whom Hitler would negotiate any peace terms, the logical director of England's destiny after the war."

While there are both German and American files about the

wartime activities of the Duke and Duchess of Windsor, only the British held a copy of a German file innocuously labeled as "Matters relating to the Royal Household 1940–66." This includes a top secret file dated August 24, 1953, containing the German correspondence relating to the Nazi plan for the Duke and Duchess of Windsor during their stay in Madrid and Lisbon during the summer of 1940. This file was recovered at Marburg Castle at the end of the war. The "Marburg Files" mostly consist of an exhaustive list of the trove of stolen paintings and decorative artwork that was hidden there. When the Marburg site was closed in 1946, the artworks were sent to Wiesbaden for identification and eventual restitution. The files pertaining to the duke and duchess were handed to the British, where they were promptly classified by Churchill. The files were reclassified periodically over the ensuing years and again as recently as 2000 under the UK Freedom of Information Act. When the American surveillance files were slated for publication in the early 1950s, Churchill again intervened with President Eisenhower to suppress their publication.

The Marburg Files reveal in detail the vulnerability of the duke and duchess to the Third Reich and the perilous state of British democracy as a result. The episode began with a secret cipher from Ribbentrop dated June 24, 1940, inquiring if it was "possible to keep the Duke and Duchess of Windsor in Spain for at least a few weeks before granting them a further exit visa? It would of course be necessary that it should not in any way leak out that the suggestion comes from Germany." The German ambassador, Hoyningen-Huene, cabled Ribbentrop to tell him that the duke "is convinced that had he remained on the throne, war could have been avoided," describing himself as a "firm supporter of a peaceful compromise with Germany. The Duke believes with certainty that continued heavy bombarding will make England ready for peace."

Ribbentrop also engaged the German ambassador to Spain, Stohrer, in a cable to be kept "under lock and key," that "the Duke must be informed at the appropriate time in Spain that Germany on her side wishes for peace with the British people, that the Churchill clique is standing in the way of this, and that it would be a good thing if the Duke would hold himself in readiness for further developments. Germany is determined to compel England to make peace by the use of all methods and would be prepared in such an event to pave the way to the granting of any wish expressed by the Duke, in particular with respect to the ascension of the English throne by the Duke and Duchess." According to the agent who communicated this offer to the duke and duchess, they did not at first understand the Nazi offer, pointing out that a return to the throne was impossible according to the British constitution following an abdication. The same Nazi go-between "then remarked that the course of the war may produce changes even in the British constitution," at which point "the Duchess in particular became very thoughtful."

Both Churchill and the English intelligence services followed the royal couple's migration to Spain and understood the duke's vulnerability to Ribbentrop's plan, especially after having sought refuge in Madrid, as well as learning of the couple's role in passing strategic government information to the enemy. Lord Caldecote, secretary of state for Dominion Affairs, petitioned Churchill directly to send the duke and duchess to the Bahamas for the duration of the war where they would not have access to sensitive government and military information. He wrote to Churchill that "the activities of the Duke of Windsor on the Continent in recent months have been causing HM and myself grave uneasiness as his inclinations are well known to be pro-Nazi and he may become a centre of intrigue. We regard it as a real danger that he should move freely on the Continent. Even if he were willing to return to this country his presence here would be most embarrassing both to HM and the Government."

When Churchill asked the duke and duchess to return to England, they declined. In response, Churchill insisted that any refusal to follow the orders of the British government would be considered a serious breach and urged the duke to comply with government wishes to go to the Bahamas. In addition to a swiftly imposed departure from Europe with the appointment of the duke as the governor of the Bahamas, the British government also enforced a news embargo on the royal couple in both local and international media. Ribbentrop noted that according to a Swiss agent with close ties to the British Secret Service, the intent in sending the duke to the Bahamas was "to do away with him at first opportunity."

The Nazi high command in Spain informed Berlin in a series of top secret cables that "Windsor told the Foreign Minister that he would only return to England if his wife were recognized as a member of the Royal Family and if he were given an influential post of a military or civil nature. Fulfilment of these conditions was as good as impossible." After this unsatisfactory ultimatum was presented to the duke, the Germans planned to notify him about the threat to his life, with an offer from Spain, as negotiated by Ribbentrop, "to accept Spanish hospitality and if necessary financial assistance," on behalf of Germany.

Ribbentrop followed developments in Madrid and then Lisbon, where the duke and duchess were the houseguests of a pro-German Portuguese banker, laying the foundations for the royal couple to remain in Spain. He even vetted the plan with the Spanish minister of the interior, the brother-in-law of the "Generalissimo" who eventually acted as an intermediary between the Nazi high command and the duke. Ribbentrop noted that Churchill had issued a verbal ultimatum to the duke threatening court-martial if the couple did not leave Europe for the Bahamas immediately.

While secret cables crisscrossed Europe updating the Reich on overtures to the royal couple, Ribbentrop also sent an agent to Lisbon to speak with the duke and duchess about an attractive alternative offer made by Ambassador Stohrer on their behalf. In a lengthy cable to Ribbentrop marked "Urgent and Top Secret," the ambassador summarized two meetings in Lisbon, one with the duke alone and a second meeting with both the duke and duchess. Stohrer noted that at both meetings, the "Duke expressed himself most freely," elaborating on his interest to "break with his brother" the king, whom he finds to be "altogether stupid," and with "the clever Queen," whom he believed was conspiring against both himself and Wallis. "The Duke and Duchess declared that they would very much like to return to Spain, and expressed thanks for offer of hospitality." Stohrer concluded with the recommendation that the duke and duchess should "leave Lisbon in a car as if he were going on a fairly long pleasure jaunt, and then to cross the border at a specified place, where Spanish secret police will ensure a safe crossing." In a follow-up cable to Ribbentrop the following day, he confirmed that the "Duke and Duchess are prepared to return to Spain."

When the Nazis presented the duke with an official letter containing the details of their plan to return the royal couple to Madrid, the duke "became very thoughtful but finally only remarked that he must consider the matter; he would give an answer in 48 hours." After receiving word that the duke and duchess were planning to depart for the Bahamas regardless of the German offer, Ribbentrop again implored his agent in Lisbon, in a telegram to be placed "personally under lock and key," to tell the duke again that "Germany would be ready to cooperate closely with the Duke and prepare the way for the fulfilment of every wish expressed by the Duke and Duchess." The telegram goes on, "[I]f the Duke and Duchess have other intentions, but would be ready to cooperate in the restoration of good relations between Germany and England at a later date, Germany is equally ready to collaborate with the Duke and shape the future of Duke and Duchess in accordance with their wishes."

The German cables became more urgent as news of a goodbye party for the duke and duchess at a Lisbon hotel, as well as imminent sailing dates for the Bahamas, became known. The duke offered as an excuse that England was not yet ready for peace and that, therefore, the job of reconciliation with Germany would not succeed at his initiative alone. He proposed to maintain contact with a German agent from the Bahamas, agreeing on a code word that would initiate his return to Europe aided by the Nazis. It was also noted that as part of this discussion, the duke "expressed admiration and sympathy for the Fuehrer."

While all indications were that the duke planned to leave for the Bahamas, the German high command, in a lengthy "Urgent" telegram "to the German Foreign Minister Personally," set out a detailed contingency plan to spirit the duke and duchess safely from Lisbon to Nazi-controlled Spain as well as last efforts to "aggravate a motive of fear and persuade the Duke and Duchess to remain in Europe." Among the plans were to fire shots near the couple's Lisbon hotel, an idea that might backfire and encourage them to leave for a safer place, or a bouquet of flowers delivered to the duchess with a note of warning about the perils that awaited them in the Bahamas.

The Nazi high command believed that the strongest argument to coerce the Duke of Windsor to decide to officially help Germany would be to play on the duke's anti-Semitism: "as the Duke is particularly impressed by the Jewish danger, a list was handed to his private secretary, Philipps, of the Jews and emigrants travelling on the same ship, emphasizing the fact that Security Police could offer no guarantees." In the end, the Windsors boarded the steamship *Excalibur* to the Bahamas, wavering on the decision until the very last minute, even causing a delay in the ship's departure, but persuaded one last time by Sir Walter Monckton, who had traveled on behalf of Churchill to Lisbon to ensure that they left.

The German high command believed that the Duke of Windsor departed with the conviction that he would still be able to intervene on behalf of Germany from his post in the Caribbean. A secret telegram to Ribbentrop reported that two weeks after their departure, our "Agent has just received a telegram from the Duke in Bermuda asking him to let him know as soon as action is necessary on his part. Shall any answer be sent?" This was the last communiqué on this topic.

The American intelligence file on the Windsors was dormant until June 1953, when a secret and personal letter from Churchill to Eisenhower was added as an addendum. Starting with "My dear Friend," Churchill goes on to refer to a microfilm of the German telegrams that were found in German archives and were about to be published in the United States as part of an official history of the war. In his letter, dated some thirteen years after those events, Churchill warned that if this file of telegrams between the Nazis and their agents in Spain and Portugal "were to be included in an official publication they might leave the impression that the Duke was in close touch with German Agents and was listening to suggestions that were disloyal." Churchill said he had convinced the king to appoint the duke and duchess to a post in the Bahamas to prevent their entrapment by the Germans and even potentially a plan to kidnap the duke. Churchill ends the letter saying that he

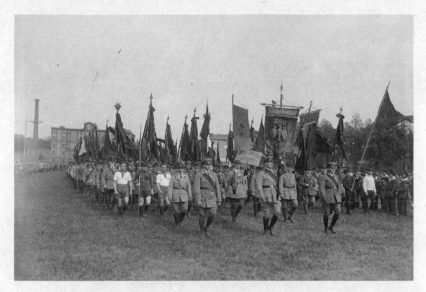

The Reichsbanner Paramilitary, Berlin, 1928
(Copyright © Bundesarchiv)

Poster for the film *Der Ewige Jude* (The Eternal Jew), 1940 (Copyright © US Holocaust Memorial Museum, courtesy of Museum für Deutsche Geschichte)

Oswald Mosley saluting a brigade of female
Blackshirts, London, mid-1930s
(Copyright © Sueddeutsche Zeitung Photo/Alamy Stock Photo)

British Union of Fascists' cover of *Action*, one of many
anti-Semitic newspaper headlines, January 14, 1939
(Copyright © The National Archives)

The young Queen Elizabeth giving the Nazi salute with Queen Mary, the
Duke of Windsor, and Princess Margaret, c. 1933
(Copyright © The Sun/Alamy Stock Photo)

Portrait of Nancy Astor
painted by John Singer
Sargent, 1908
(Cliveden Estate,
Buckinghamshire)

Cliveden House

Putzi Hanfstaengl
consulting with Heinrich
Himmler, 1931

Unity Mitford as Hitler's guest at the Bayreuth Festival, 1936

(Copyright © Pictorial Press Ltd./Alamy Stock Photo)

The Mitford sisters: Unity (*left*), Diana, and Nancy

(Copyright © Mary Evans/Marx Memorial Library)

The Duke and Duchess of Windsor greeted by Hitler on their visit to Germany, 1937
(Copyright © Pictorial Press Ltd./ Alamy Stock Photo)

Adolf Hitler and the Duke and Duchess of Windsor during their visit to Berghof in Berchtesgaden, October 22, 1937
(Unknown photographer)

Anna Wolkoff, convicted of spying for the German government
(Copyright © Hulton-Deutsch Collection/Corbis via Getty Images)

Mug shot of Tyler Kent, Anna Wolkoff's convicted accomplice in spying for the German government
(Copyright © The National Archives)

Unity Mitford being repatriated from Germany at
Folkestone Harbor, January 1940
(Copyright © British Pathé/Mary Evans)

Parish church of St. Giles, Wigginton, the location of the maternity home
where, according to local accounts, Unity gave birth in the spring of 1940
(Copyright © Ben Nicholson)

intends to ask the French government to support his request to "prevent the publication" of this information.

Eisenhower responded about a week later in a cable marked secret and personal that he had been aware of the files since they were recovered by the US military in 1945 and agreed that "there was no possible value to them." While he expressed surprise that the files had been converted to microfilm, Eisenhower said that he was not sure what, if any, classification they might have, or whether, as a result, he would be able to control their publication, something he promised to look into. A month later, former British prime minister Clement Attlee also wrote to Churchill agreeing that the "publication of these documents might do the greatest possible harm," adding that he had made a similar request to Eisenhower to bar their publication. Churchill responded to Atlee's letter quickly the following day that "I earnestly trust it may be possible to destroy all traces of these German intrigues." While Churchill also wrote to the French foreign minister to enlist his support, there is no response in the file, which ends abruptly. In the end Churchill did not succeed in having the files reclassified, or the book's publication suppressed.

The Nazi high command was not alone in keeping a file about the Duke and Duchess of Windsor. The couple also captured the attention of the American intelligence community, likely because in their Bahamas exile, they were only a stone's throw and a security risk to the US mainland. At the request of President Roosevelt, the FBI opened intermittent investigations and surveillance of the couple, dating to their arrival in the Bahamas in 1940 and corroborating a shared interest in the couple's recruitment by Ribbentrop. The FBI noted that the duke claimed to have only known Ribbentrop "in his official capacity and never saw him after 1937." The FBI files tell a different story, one that was explosive enough to be actively suppressed by Churchill and repeatedly reclassified by the United States.

The FBI file reflects comprehensive surveillance initiated by President Roosevelt and overseen by the FBI director, J. Edgar Hoover, that confirms much of the Nazi correspondence in the Marburg Files, with an emphasis on the duchess's duplicity and influence on her husband. In a confidential memo to Hoover, the lead agent on the case writes that "it has been ascertained that for some time the British Government has known that the Duchess of Windsor was exceedingly pro-German in her sympathies and connections and there is strong reason to believe that this is the reason why she was considered so obnoxious to the British Government that they refused to permit Edward to marry her and maintain the throne." It goes on to state that the duke "is in such a state of intoxication most of the time that he is virtually non compos mentis." The duchess had maintained "constant contact" with von Ribbentrop and because of the duke's position, she was privy to sensitive information regarding both the British and French governments, which she didn't hesitate to send along to the German high command.

According to the FBI report, the British government was "always fearful that the Duchess will do or say something which will indicate her Nazi sympathies and support, and consequently it was considered absolutely essential that the Windsors be removed to a point where they would do absolutely no harm." The governor of the Bahamas was removed within two hours and a plan put in place to send the couple to the farthest and most isolated part of the British Commonwealth, with extra care put in place to ensure that the duchess had no communication with Ribbentrop.

The FBI began investigating the duke and duchess as soon as they arrived in the Bahamas. A memo in the FBI file dated October 1940 flagged that the duchess, "violently pro-German," was sending her clothes from the Bahamas to a dry cleaner in New York City: "the possibility arises that the transferring of messages through the clothes may be taking place." Another lengthier report detailed a tip from a female informant, name redacted, who was affiliated with the USO Club of Cincinnati. She said that the duchess was "an associate of von Ribbentrop" and additionally, she was acquainted with a Scottish reverend named McCloud whose son was a missionary in East London near Limehouse, where the British Union of Fascists was active. The reverend's son said that drugs were sold across the street from the mission where he worked and that he had "observed the Duchess of Windsor enter one of these places to purchase 'dope,' which she supplied to the Duke, earning her the title of 'Limehouse Bess' among the British aristocracy."

The American file also includes a 1942 interview with a Benedictine monk in Boston, underscoring the lengths to which the FBI went in order to conduct surveillance on the duke of Windsor.

Father Odo, the former German-born Carl Alexander, Duke of Württemberg, believed when the FBI arrived that he was being questioned about his friendship with Queen Mary. Only later did he realize that the Duke of Windsor was the real target of the interview. Father Odo told the FBI that he had never met the Duke of Windsor but had "seen the Duchess of Windsor on numerous occasions and is aware that she had formerly consorted with von Ribbentrop. . . . Father Odo further confided that he knew that von Ribbentrop, while in England, sent the Duchess seventeen carnations every day. The seventeen supposedly represented the number of times they had slept together."

A second FBI interview with Father Odo further confirmed the complicity of the royal family in covering up the duke and duchess's pro-Nazi agenda. According to FBI Agent Laughlin, Father Odo said that Queen Mary had told him that Wallis was Ribbentrop's mistress and that the royal family forced him to abdicate "because of the scandal which he had brought upon his country by associating with a mistress, and secondly because of his fondness for Nazi ideology and his desire to rule England without a parliament. . . . The Duke wanted to suppress Parliament and head a party which in effect would make him the Dictator of England." The FBI file of Father Odo's interviews were immediately classified and reclassified in 1980 and again in 1990.

The American files also lay out in detail the case against the Duchess of Windsor, who, according to FBI sources, was a well-known German intelligence asset. One report, marked "Confidential," references a source that claimed that Wallis "was being maintained and financed by von Ribbentrop and was on the pay-roll of the German Government at the time of her marriage . . . AND THAT IS THE REASON THAT EDWARD WAS NOT ALLOWED TO TAKE THE THRONE OF ENGLAND."

Another FBI file details that Ribbentrop recruited Wallis Simpson early on and paid her handsomely to furnish the Nazi government with information about the British "social set." It claims that Hitler and Ribbentrop together brainstormed the idea of her marriage to Edward, who was already showing interest in their cause. They believed that Wallis "will see to it that Edward will never sign a declaration of war against Germany," a plan that was scuttled by his unforeseen abdication, which also denied her the opportunity to be Queen. As "revenge, she took her husband the Duke of Windsor to Berlin to see Hitler and there was a deal made between Hitler and the ex King." The deal outlined was that the duke would use his influence with both the British military and the royal family to ensure that Britain did not resist a German invasion. In exchange, Hitler promised to install the duke as king of England after the war. The Nazi high command believed "this was easy for the Duke as many British high officials still think that he is their legal king and the British anyhow are not anxious to help the Russian Communists." It goes on to explain that Hitler informed the Japanese of the duke's role and urged them not to attack the United States since the British would not resist invasion, citing the evidence of the British withdrawal from Hong Kong, Singapore, Burma, and Libya. The American intelligence assessment also detailed that "there is no question about it that Churchill is being double crossed" and that through the duke, the duchess is privy to all sensitive communications between the British and Americans—and passes the information to Hitler. It closes with the warning that "she may succeed and if she does it will cost the lives of millions of Americans."

While Wallis represented one line of inquiry, the FBI turned its rigorous attention to the duke as well. In a confidential report dated April 1941 and marked "destroyed" in April 1960, an agent wrote directly to Hoover reporting on a debrief from an informant, William Rhinelander Stewart, who had visited with the duke and duchess in Nassau. Stewart relayed that the duke regarded Nassau as his exile in Elba and confirmed "that there was current rumor and gossip in Nassau to the effect that when Hitler defeated England, he would then install the Duke of Windsor as the king and that, of course, under such circumstances there would be no question about his wife's being queen." Only under a Nazi regime, would she finally have that chance.

The proposed six-day trip by the duke and duchess to Florida in April 1941 for a dental visit triggered scrutiny by the FBI at the highest ranks of the organization. A confidential memo to FBI director J. Edgar Hoover laid out the case for the FBI to add a clandestine surveillance unit in addition to the security detail given to the duke and duchess for their short visit. While the British government sought to isolate and suppress publication of the information about the duke and duchess, the American government sought to actively investigate it.

In anticipation of their trip and in light of the proximity of their

Nazi connections, the assistant secretary of state, A. A. Berle Jr., wrote to the attorney general, Mr. Alexander Holtzoff, requesting "discreet observation . . . and a wider and less obvious cover" during the Windsors' visit to Miami and Palm Beach. In response, Hoover made it clear that this request was being made at the suggestion of the president and would be further cleared by the Treasury Department, Secret Service, and other individuals "watching the Duke and Duchess." Later classified correspondence underlines the president's request for better surveillance of the duke, suggesting that was difficult to achieve until an agent proposed asking a British friend of the duke who was trying to become a naturalized US citizen to serve as a go-between and introduce an FBI undercover agent into the duke's Palm Beach golf foursome as a personal friend. This plan worked well and provided additional intelligence that the duke was golfing with Latham Reed who had given a speech that had "praised the German government and the Hitler regime."

Once again, the duke and duchess slipped through whatever surveillance was put in place, departing Palm Beach on April 23 at 10:30 A.M. on a private plane belonging to William K. Vanderbilt. The duke and duchess returned to the United States one more time before departing from the Bahamas, with no surveillance.

After nearly five years as governor of the Bahamas, the duke resigned his post in March 1945, marking the end of the war, the defeat of the Nazi regime—and ostensibly, the couple's liability to the British government. The duke announced that his "resignation does not mean a permanent severance from public life because after the war men with experience will be badly needed and I'll fit in anywhere that I can be helpful." British newspapers speculated that rather than return to Britain, the duke and duchess would travel to the United States, perhaps visit their ranch in Canada, and ultimately return to their homes in Paris and Cap d'Antibes in the South of France.

At the end of the war, the Duke of Windsor asked to tour the FBI facilities at Quantico, Virginia. This tour was personally undertaken by J. Edgar Hoover, each moment carefully choreographed, with no reference to the fact that the duke was previously the subject of an intensive FBI investigation. The duke wrote a note of thanks for the tour, making a point of mentioning how impressed he was "by the scope of your activities." In a short note, the duke also mentioned that he was in close touch with Captain Guy Liddell, the lead MI5 investigator during Unity Mitford's repatriation from Nazi Germany and part of the cohort of the Cambridge Five group of double-crossing spies on behalf of the Soviet Union. The note ended without irony, claiming that "personally, I should like to say how pleasant it was to renew my acquaintance with you."

Joachim von Ribbentrop was arrested at the end of the war in June 1945 and was among the first to be tried at Nuremberg for his role in "the destruction of the European people." The charges against him dated to 1932, including his time as ambassador to England. At his trial, Ribbentrop confirmed that he traveled to England as early as June 1935 at Hitler's request to seek out influential Britons who would be interested in a German-British pact, and that he advanced that possibility with the many highly placed Britons, including politicians and aristocrats, he met in his capacity as ambassador.

Ribbentrop testified at Nuremberg that he believed that he had been close to reaching an even more comprehensive agreement than the Anglo-German Naval Agreement (1935) with the Chamberlain government. To this end, he had breakfast with Prime Minister Chamberlain at 10 Downing Street at which Chamberlain "emphasized his desire to reach an understanding with Germany. I was extremely happy to hear this and told him that I was firmly convinced that this was also the Fuehrer's attitude. He gave me a special message for the Fuehrer that this was his desire and that he would do everything he could in this direction." Subsequent meetings to work on the details followed, including an invitation to Foreign Minister Lord Halifax to return to Germany to conclude the final details with the promise that "another exhibition of hunting trophies could be arranged." Halifax was an avid hunter and this exhibition offered a pretext for the real purpose of the trip, to discuss the matter with Hitler at Berchtesgaden. Ribbentrop concluded that while there were "many Englishmen who had a very positive attitude towards this idea," it was one of Hitler's great disappointments that the "policy did not materialize."

Ribbentrop's reach into the halls of power in English government and high society is revealed by the list of prominent witnesses he attempted to call to his defense during the trial, including Prime Minister Winston Churchill. The list was so long and potentially incriminating that Sir Alexander Cadogan, the permanent undersecretary of state for foreign affairs and principal landowner in Central London, sent a classified letter to Churchill informing him that the list had already been shared with President Truman and Marshal Stalin. While Churchill downplayed the legitimacy of the witnesses he called, Ribbentrop's testimony was long and detailed. He was found guilty and hanged at Nuremberg at 1:30 A.M. on October 16, 1946. In a group that included Julius Streicher, he was among the first defendants to be executed for his part in the Nazi regime. Both Hermann Goering and Martin Bormann were supposed to be part of the group but committed suicide first.

The Duke and Duchess of Windsor never returned to Britain. After the war, they lived in a villa in the Bois de Boulogne in Paris. Their neighbors and lifelong friends were Oswald Mosley and his wife, the former Diana Mitford.

CHAPTER 7

THE WOLKOFF AFFAIR

Soon after Adolf Hitler was designated *Time* magazine's Man of the Year (1938), Joseph Kennedy, American ambassador to the Court of St. James's and scion of the Kennedy family, stated in an interview, "Democracy is finished in England. It may be here (in the US) too." Britain's relationship to America was central during this period. Churchill believed that unless Britain entered the war with American support, the Commonwealth could not survive and that the entire European bloc, including Britain, would fall to the Nazis, possibly for generations to come. Roosevelt was constrained not only by public opinion in favor of isolationism but by Kennedy's predictions of Britain's swift defeat by Germany, warning that any American military aid would end up in Nazi hands. Churchill's private correspondence with Roosevelt, written under the pseudonym "Former Naval Person," was stolen by Nazi sympathizers in England and handed to the Reich late in the 1930s.

While Ambassador Kennedy arguably harbored some sympathy for Germany, other forces were at work in the American embassy to undermine Churchill's pleas for American intervention. Anna Wolkoff, a White Russian émigré in London and secretary of the Right Club who operated out of a tearoom in Kensington, recruited Tyler Kent, a low-level employee at the American embassy, to steal secret correspondence between Roosevelt and Churchill requesting American military aid. Both Wolkoff and Kent were eventually arrested and indicted.

The trial of Anna Wolkoff and Tyler Kent, heralded by front-page headlines in the British press as "the biggest trial of the war," took place in a sealed courtroom at the Old Bailey in October 1940. Tried separately, Wolkoff and Kent were accused of stealing top secret cables exposing both Britain's strategic military weaknesses and Roosevelt's willingness to support England, thereby violating American neutrality. The contents of these documents were so inflammatory, especially to the incumbent American president, that Churchill tried unsuccessfully to have the trial postponed until after the US presidential election in early November 1940. Kent was also refused extradition and tried as a war criminal in England.

Despite Churchill's repeated efforts, the Wolkoff trial commenced on October 23, 1940. In early November, Roosevelt won the presidency in a landslide, protected by the cloak of secrecy about the issues at stake and assuring that none of the evidence fell into the hands of the press. Describing security at the Central Criminal Court, the *Daily Express* reported that "thick brown paper was pasted over the glass panels of the doors, the doors themselves were locked, and police stood guard at them to ensure that nobody outside the courtroom would see a certain witness." The British press speculated correctly that the star witness at the trial was likely to be high-ranking MI5 agent Maxwell Knight.

The accused Anna Wolkoff was a dressmaker and sole proprietor of Anna de Wolkoff Haute Couture Modes. Her atelier was located on an especially upscale stretch of Conduit Street in Mayfair, home of London's high-end retail boutiques. Despite clientele like the Duchess of Windsor, Wolkoff's business barely broke even and eventually closed when the principal backer withdrew, leaving Wolkoff with £4,500 in debt. A naturalized British citizen, Wolkoff was also an active member of British right-wing movements during the 1930s. Her Fascist leanings were influenced by her Russian past. She was rumored to be terrified of the plight of General Kutepov, a white Russian war hero who was kidnapped in Paris and sent to a Russian prison where he was tortured to reveal the names of White Russian agents living abroad who had turned against Stalin. He was never heard from again, and Wolkoff feared a similar fate.

As part of a pro-Nazi European network, Anna Wolkoff was protected by European aristocrats and members of government. The information in one of Wolkoff's personal intelligence files was considered so damaging that it was initially classified for one hundred years, until 2044, and only recently published under the Freedom of Information Act.

Wolkoff's co-conspirator, Tyler Kent, was a twenty-nine-year-old cypher clerk at the American embassy in London, where he mostly worked the 4:00 P.M.-to-midnight shift in the embassy code room. A prep school and Princeton graduate, Kent's first diplomatic posting was to the American embassy in Moscow where he had a reputation as more of a ladies' man than a diplomat, forming a romantic liaison with an agent in the Russian Narodny Komissariat Vnutrennikh Del (NKVD), forerunner of the KGB. Once in London, Kent's political views attracted the attention of the local police who initiated surveillance, quickly concluding that he was both anti-British and pro-German. Despite his limited salary, Kent maintained bank accounts in London, Cork, and Dublin and used a few different addresses in Chelsea and Knightsbridge. He also traveled frequently outside the UK, which further raised the suspicions of British intelligence.

Wolkoff met Tyler Kent at the Right Club where both were outspoken members. Wolkoff recognized a fellow sympathizer in Kent, to whom she spoke in Russian under the pretext that "it is very convenient to be able to talk in rapid Russian, because if the lines are tapped they will not be able to understand." An informer who worked in the British government helped Wolkoff disseminate the top secret correspondence between Roosevelt and Churchill to both the Nazi Party and Nazi sympathizers across Europe.

Maxwell Knight, a renowned British spy master who later became Sir Maxwell Knight and the model for the James Bond character "M," was an ambitious young officer in MI5. Knight reported directly to Guy Liddell. Knight had established a secret office in Dolphin Square in Pimlico and recruited Guy Burgess, who along with Liddell's other later recruits, Anthony Blunt and Kim Philby, as well as Donald Maclean, became known as the Cambridge Five spy ring that passed sensitive British government intelligence to the Soviet Union during the war and well into the 1950s.

From an unremarkable flat in Dolphin Square, a location chosen because it was also home to Sir Oswald Mosley and Diana Mitford, Knight and Burgess infiltrated the Right Club and quickly focused on Anna Wolkoff. Among an increasingly persuasive and vocal group of German sympathizers, Wolkoff was an early and forceful convert to the Nazi cause. She visited the Sudetenland just before its annexation by Germany and boasted of her close ties to the Nazi leadership and their open invitation to visit whenever and whatever she wished.

Wolkoff's subversive activities outside of the club quickly gained Knight's attention. He recruited informants within the Right Club, which led him to file an affidavit against Archibald Ramsay and his wife, the widow of Lord Ninian Crichton-Stuart. When the Ramsays were not living at the medieval Kellie Castle in Scotland, they maintained an apartment in Ovington Square in London where they recruited new Right Club members like Anna Wolkoff. Testimony by a member of the Christian Protest movement detailed how a "political and religious" movement worked to establish close ties between Christian conservatives and the Far Right. That witness recounts that Mrs. Ramsay, or Ismay as she was known to club members, liked to boast that she had met Hitler personally as part of a British delegation and that "Hitler was undoubtedly a genius."

Another informant activated by Maxwell Knight was Anne Van Lennep. A British citizen of Dutch background, she maintained a close friendship with Baroness Fuchs von Norhof who left England on the eve of the war in 1939 suspected of being a Nazi spy and was "very closely associated with Anna Wolkoff." Lennep told Knight that Wolkoff had "made herself very useful" and that "when the crisis comes," the Far Right had contacts in both MI5 and the police. Another of Knight's informants, "Mrs. X," whose name is redacted to this day, told Knight that "many of the names of the Right Club did not appear in any written record. The ones that did were kept in a special locked book, so that if I joined I need not fear that my connection with the movement would become known. This left me with the impression that membership of the Right Club was a secret matter."

While Wolkoff was an active member of the Right Club, she operated her other subterfuge activities out of the Russian Tea Rooms on Harrington Road in South Kensington. Knight focused his surveillance on activities of this White Russian émigré gathering spot in order to understand and infiltrate an insurgent cell of Nazi support.

The café was composed of two shabby dining rooms, one

hung with an imposing portrait of the last tsar. The Russian Tea Rooms had been established by Wolkoff's parents in 1923 and was a frequent gathering spot for members of the club where they could be assured that their tea was served with appropriately subversive pro-Hitler sentiments. One tearoom patron, a naturalized Belgian, recalled that "there was a group of people who met there very often. They generally sat together at the same table and talked in low voices. The group consisted mainly of women, but occasionally there was a man or two. The principal topic of their conversation was anti-Semitism and praise of Germany for the way it had rid itself of Jews."

Admiral Nikolai Wolkoff, Anna Wolkoff's father and owner of the Russian Tea Rooms, also attracted the attention of MI5. Wolkoff was a former military officer who spoke frequently about his interest in the Balkans and Russia. His intelligence file described him as forty-five to fifty years old, about five feet, seven inches tall, clean-shaven with "dull blond hair; grey blue eyes; reddish complexion; hunched walk, wears a gold-rimmed monocle." On the same day that Mrs. Hall, a Knight informant, lunched at the Russian Tea Rooms, Admiral Wolkoff was overheard describing the German invasion of England and how he "hoped nothing would happen to him because he wanted to see the day when he would spit in the faces of the English."

Knight also interviewed Lieutenant Nicholas Ignatieff, a member of the First Canadian Division of Engineers and eldest son of Count Pavel ("Paul") Ignatieff, minister of education in the Imperial Russian War Cabinet of 1914. The Ignatieff family had escaped to Canada three years after the Revolution where Ignatieff's father maintained close connections to the émigré community as the president of the White Russian Red Cross Society. The younger Ignatieff, who lived in London, explained that its White Russian community was "quite fanatical in its anti-Semitism and really have come to regard Hitler as a savior, not only of Germany, but of the whole of Europe including Russia. They will do literally anything to help his cause, so ardent is their belief in the new order which they think he is to bring to Europe and possibly the world." Lieutenant Ignatieff went on to say that "the Nazi movement was beginning to make a good deal of headway in Canada until the outbreak of the war."

Sir Archibald Ramsay and his wife lived down the street from the Russian Tea Rooms and closely monitored Wolkoff's contributions in furthering the Nazi cause. Tyler Kent, whose dalliances with the White Russian émigré clientele presented an opportunity for him to both practice his Russian and further his pro-German ties, was also a tearoom regular. Other members of the Right Club who spent time at the tearoom and became subjects of interest to MI5 included Lady Beatrice Kerr-Clark, who lived nearby and described Anna Wolkoff as "a very personable type of woman with many man friends." She added that the Russian Tea Rooms was "frequented by persons of a violently anti-Semitic type. A conversation in the tea shop has been overheard by a friend of Lady Beatrice to the effect that 'it would be a good thing if Hitler won.'"

Knight's surveillance also focused on various other patrons.

From the MI5 files: "Alan Wilfred Devon George, of 18 Queens Gate Place, S.W., a pro-Nazi, is very friendly with Admiral Wolkoff and his two daughters and until about a fortnight ago, dined at the Russian Tea Rooms in Harrington Gardens almost every night." Mrs. Austin Hall who had lunch at the café on June 28, 1940, was observed to live in South Kensington with her young son and a German nurse. "Mrs. Hall was pro-German in her attitude, though perhaps not definitely pro-Nazi." She was overheard to be speaking about a friend, Mrs. Arthur Watson, who had been interned for seven weeks at the Royal Holloway Prison. She also boasted that she corresponded frequently with Graf (Count) Montgelas and sent him one pound weekly to his internment camp, in addition to other pro-Nazi Austrians living in Central London with whom she was in frequent contact.

A statement by Joan Priscilla Miller, a secretary at the Military Intelligence Department, met Wolkoff at the Russian Tea Rooms. She described "Anna Wolkoff's conversation (as) strongly anti-Semitic. She said she had had to close down her dressmaking business because of the unscrupulous tactics of her Jewish competitors and that the Jews were ruining the country." Wolkoff recognized that Miller's job in the British government might present an opportunity down the line and sent her a dress of Wolkoff's own design, left "lying on her doormat," that Miller had admired. At another meeting, MI5 notes that the two women made omelets together. Miller recounted that Wolkoff talked a lot during that dinner and "said she had been listening in to the news on the

German radio and expressed the opinion that the German military and air strength was far superior to ours and that England would not be able to stand up to Germany."

Anna Wolkoff capitalized on a budding friendship with Tyler Kent to persuade him to bring home sensitive documents with information about US and British strategy that would benefit the Nazi regime. Wolkoff would stop by Kent's flat and "borrow" the documents, after which she would have them photographed by family friend and amateur photographer Nicholas Smirnoff. Like Wolkoff, Smirnoff was a White Russian by birth and a naturalized British citizen. He worked as an examiner in the British Censorship Bureau until he was eventually detained. Wolkoff told Smirnoff that "the telegrams were very confidential and the information in them was wanted for her society." In a sworn statement, Smirnoff said that Wolkoff "told me that these documents were lent to her by a foreign diplomat, but did not say who he was or from what Embassy he came. She said it was a stroke of luck that she had managed to get these documents and that in these documents preference had been given to the United States—I cannot say what for. . . . I thought it was rather strange that she had these telegrams in her possession. I could not read what was on the pictures I had made, but Anna Wolkoff said it would not matter because it could be read with a magnifying glass."

According to the prosecution at his trial, a raid on Kent's flat "found masses of highly confidential cables, codes and copies of such documents, many of them extremely confidential and of the greatest importance to the diplomatic and strategic position of the Allies." There was at least a cursory concern on the part of the American embassy that Kent had not acted alone in supplying Wolkoff with the cables. Tevis Wilson, also employed in the cipher and code department, was a close personal friend of Kent's and was quickly moved to the visa section of the embassy following Kent's arrest, with no further action taken.

The telegrams stolen by Kent and passed to Wolkoff contained information about the decrepit state of the British fleet and urgent requests by Churchill for military support from America. The trove included two top secret messages from Churchill to Roosevelt. Disclosures included instructions to the British Fleet on how they should handle American merchant vessels and proof that Churchill, then serving as First Lord of the Admiralty, was eschewing official protocol and bypassing the prime minister to go directly to Roosevelt for military aid aimed to lure the United States into the war, which Churchill believed was a foregone conclusion. In cables marked STRICTLY PERSONAL AND CONFIDENTIAL, FOR THE PRESIDENT, MOST SECRET AND PERSONAL, Churchill pleaded for help from Roosevelt: "Immediate needs are: first of all, the loan of forty or fifty of your older destroyers to bridge the gap between what we have now and the large new construction we put in hand at the beginning of the war. This time next year we shall have plenty . . . anti-aircraft equipment and ammunition, of which again, there will be plenty next year if we are alive to see it. . . . We shall go on paying dollars as long as we can, but I should like to feel reasonably sure that when we can pay no more you will give us the stuff all the same."

Wolkoff passed the documents she "borrowed" from Kent to a fellow Nazi sympathizer, a staff member at the Romanian embassy, who expedited the Kent telegrams by diplomatic post to another Right Club member, William Joyce. Joyce, referred to in a popular song as "Lord Haw-Haw the humbug of Hamburg," had left England to establish a pro-Nazi English-speaking radio show in Berlin which he started in early 1940. According to British intelligence, Joyce's show "broadcasts the most virulent fifth column propaganda and at times gives instructions to secret groups or cells in the country to carry out subversive operations in Great Britain." Joyce's program appeared on frequencies across Europe, not just in Britain. The documents Wolkoff received from Kent, often written in code, described, among other things, Jewish activities in England. Wolkoff thought Joyce could make use of these in his propaganda broadcasts from Germany and that the information would "be like a bombshell." On this particular occasion, a trap was set and the contact unknowingly handed the document to Maxwell Knight of MI5 instead of Joyce.

In addition to contacts in Nazi Germany, Wolkoff had a network of fellow Nazi sympathizers all over Europe who were eager recipients of the photos of the purloined cables, which were bundled with copies of Hitler's speech dated April 28, 1939, and pro-Nazi flyers for distribution. Wolkoff liked to claim that she had agents stationed not only in England but all over Europe and the United States, mostly female recruits she had met at Right Club meetings. Wolkoff also recruited foreign government officials such as the Count and Countess de Laudespins. The count, who worked in the Ministry of Foreign Affairs in Brussels, allowed Wolkoff to use the Belgian diplomatic post to send documents including anti-Jewish and anti-Freemason propaganda materials to be distributed to other contacts in Belgium. She ignored a warning by another Belgian contact to "give up this so-called anti-Jewish work . . . as it was dangerous." Despite the warnings, Wolkoff forged ahead, disseminating whatever information she had received to as widespread a network as she could garner, praising one go-between for doing admirable work, with "pluck and initiative."

Wolkoff also established contacts with the Italian Fascist government who had their own plan to infiltrate Britain. At his trial, Tyler Kent described how Wolkoff and a Miss Enid Riddell, another Right Club member, invited him "to a restaurant in Soho called l'Escargot, where I met a man who was introduced to me as Mr. Macaroni by Anna. It occurred to me at the time that this was probably a pseudonym. Following is a description of the man: Age about 45, shortish, thick set, dark hair and complexion"

Mr. Macaroni was really the Duca del Monte, the military attaché at the Italian embassy and a cousin of Howard Kerr, the Duke of Gloucester's equerry. Lady Howard of Effingham reported that he "looked like a cheap gangster." An early intelligence report notes that his work as attaché did not appear to be taxing and that he spent most of his time at the International Sportsmen's Club: "In spite of his ostensibly pro-British attitude, he was believed to be overstepping the legitimate duties of a military attaché in obtaining 'intelligence' in this country." There was also speculation that del Monte was not acting alone. One MI5 informant attending a cocktail party at the duke's Chelsea home at 67 Cadogan Square was suspicious that another guest, Aldo Terzolo, the head of the Italian Commercial Bank on Regent Street who was "tremendously anti-British and pro-German," might be an accomplice.

Over a convivial French meal, Wolkoff passed documents to del Monte that she had obtained from Kent and had copied at a local Woolworth's. These documents referred to "a trade agreement which benefited the United States at the expense of other neutrals" and Churchill's role in that agreement. Wolkoff "thought the Italians would be glad to have the information" and later learned through Kent that Italy was pleased to have received copies of the documents. After dinner, "Mr. Macaroni" invited the group to the Embassy Club where they stayed until just after midnight.

Miss Enid Riddell, the fourth guest at the l'Escargot dinner, described trying to contact the Duca del Monte after Wolkoff's arrest. "I told the Duca about Anna's arrest. He was very concerned and said he couldn't understand why she had been detained, nor could he understand why Kent had failed to put in an appearance. I made no further attempt to get in touch with Kent."

Eventually, MI5 trapped Kent by sending a fake cable purportedly from Roosevelt to Churchill, which, when it was passed on by the Wolkoff/Kent duo, was intended to mislead the Nazis. The ruse was actually quite simple. Kent did not notice that despite the sensitivity of the correspondence, the communiqué was sent in standard "Gray code" rather than strip cipher, an unusual format for such a high-level communication and a code that had already been cracked by the Nazis. In the fake cable, Roosevelt assured Churchill of both military and economic aid.

In addition to the cables stolen and disseminated across Europe by Wolkoff and Kent, another curious incident occurred. As MI5 began to close in on the Right Club, Sir Archibald Ramsay passed the Red Book, the secret list of the Right Club membership, to Tyler Kent for safekeeping. Kent had a history of making liberal use of US embassy safes. After the discovery of his misdeeds, US officials examined the safe at the US embassy in Moscow, where Kent had worked previously. They found a briefcase belonging to Kent that contained a revolver, a book about European languages, and pornographic pictures of his Russian girlfriend. At the trial, Kent testified that around May 1, 1940, Ramsay brought a locked "Ledger" to the embassy: "I did not know what this contained and Captain Ramsay merely asked me to keep it for him." In fact, the Red Book was purposefully left at the American embassy. A high-security Bramah lock guarded the ledger's secrets. Kent had stored it in a cupboard safe opposite the door of the cipher room discovered when the police raided his office in the American embassy. At the time, the ledger was not considered worth taking into custody along with its protector.

The value of the membership list was understood by Ramsay to be the Right Club's most carefully guarded asset. He was the only one who knew the names and identities of all the members. When Ramsay was arrested on May 23, 1940, three days after Wolkoff and Kent, he said that "he had given the Right Club membership book to Kent one night when the latter came to dinner, because he had promised certain prominent and influential members that the book would not be left in any place where any unauthorized person might have access to it." This was a promise that was kept until 1990 when the *Red Book* entered the collection of the Wiener Holocaust Library in London, one of the largest and most prestigious libraries in Europe dedicated to the study of the Holocaust and genocide.

Ramsey would boast that as a member of Parliament, "on 73 occasions . . . he had risen in his place in the House of Commons to speak on the subject of Jews and Communism. He said he was the only member who dared attack the Government on their policy and he was being removed to prison as a menace to that policy."

Wolkoff and Kent were charged under the Official Secrets Act with "offences which in substance amount either to espionage on behalf of Germany or something very closely akin to it." The OSA, as it was known, was revised in 1939 to provide official legal protection in the UK against espionage and the disclosure of unauthorized information. After Wolkoff's arrest, which was witnessed by eleven-year-old Len Deighton, who later became an acclaimed British author of spy novels, Wolkoff was taken to the Rochester Row police station, around the corner from the Russian Orthodox church where she would occasionally attend services. It was noted that a wallet with a swastika engraved on it was found among her possessions when the police raided her apartment, listed in the police report to be further evidence of her political leanings.

Kent was arrested at his flat at 47 Gloucester Place in South Kensington by Sir Maxwell Knight, who was refused entry and had to force the door open. A briefcase full of photographed secret embassy documents was found as well as a large amount of currency. When questioned by Knight at his arrest, Kent said that he took the documents because "they show a dishonest discrepancy between the news which was given to the public and the actual trend of political affairs as was known in diplomatic circles. I considered it was my duty to make these facts known." He added that he was trying out a new camera that he was thinking of purchasing from an embassy colleague who was leaving the UK, a story he later retracted when Knight visited him at Brixton Prison with proof that Smirnoff had confessed to taking the photos. In response to Kent's arrest, Ramsay commented that "T.K., poor young man, is finished; they will send him back to America because, of course, the telegrams were very secret."

From the outset of the trial, prosecution of both Wolkoff and Kent was tricky due to the nature of the crime they committed. Sir William Jowitt, the lead prosecutor, argued that "the proposed defendants take the view that they are safe from trial and punishment because neither of the governments concerned dare to publish the documents in question. It is thought they are wrong in this, and that no consideration of domestic or foreign politics need interfere with the course of justice. The extent to which communications between the two Governments can properly be disclosed in court is at present receiving active consideration by both the governments concerned." Guy Liddell, Maxwell Knight's commanding officer at MI5, would later testify that the telegrams stolen by Kent "are of such a nature that they would be of use to an enemy."

Anna Wolkoff claimed in her defense that "Tyler Kent was an agent of the OGPU (the Soviet Secret Police) and that he had beguiled Anna into doing things that she should not have done only because she was in love with him." Her defense strategy was to call witnesses, notably a Russian priest who claimed to know Kent well, to both prove that Kent was a spy—and suggest a love affair. Wolkoff's defense also noted that their client had tea the previous week with the Duchess of Kent and that "Her Royal Highness would help Anna Wolkoff to the full extent of her power."

In a sworn statement, Kent said that he had met Anna Wolkoff at a meeting of the Right Club where he was a steward, a title that he did not know the significance of. He said that he came to England to work in the coding office of the US embassy, explaining, "[M]y object in taking these was preserving for my own records which I considered of importance, without any specific object in mind." He recounted that he had struck up a friendship with Anna Wolkoff, who visited his apartment and took an interest in the documents when she saw them out, asking to borrow them. He also described an instance when he was late for a meeting with Wolkoff in his flat, and she spent an hour there alone. As further evidence of Wolkoff's knowledge of criminal intent, he claimed that when she called him from her parents' apartment on Roland Gardens, around the corner from Kent's apartment, she spoke to Kent in Russian in order to avoid potential phone taps.

Anna Wolkoff was tried separately from Kent in a sealed courtroom at the Old Bailey. Due to the sensitive political nature of the stolen documents presented as evidence, the courtroom was cleared when each count was presented. On November 7, 1940, Wolkoff was sentenced to ten years in prison for "attempting to assist the enemy," violating the Official Secrets Act of 1911 and the Defence Regulations Act of 1939. Her naturalized British citizenship was also revoked while she was in prison. The incriminating nature of Wolkoff's crimes, to both the United States and Britain, did not end with her imprisonment. In 1941, nearly a year after her trial, one of the prosecutors on the case, Bennett, filed a brief stating that he "is extremely anxious to find the third photostat in the Wolkoff case in order if possible to use it for the Prime Minister's downfall . . . if it was only a political arrangement between Rooseveldt [sic] and Churchill, it would be no use, but if it showed up malpractice it could be used as a means of dissolving Parliament."

Both Wolkoff and Kent were convicted after twenty-five minutes of deliberations. Kent was found guilty under the same Official Secrets Act as Wolkoff as well as the Larceny Act of 1916. Although they were tried separately, they were sentenced together. Wolkoff was sentenced to ten years in Royal Holloway Prison and then moved to Aylesbury Prison in Buckinghamshire. Today this prison is used for young offenders and has one of the worst ratings in the British prison system. Wolkoff's prison sentence and designation as Prisoner 352 was personally signed by the home secretary, Sir John Anderson.

With the US Justice Department declining to prosecute him,

Tyler Kent was sentenced to seven years in Brixton Prison, which he appealed. By then, his defense had evolved with Kent claiming to have taken the documents that Wolkoff had passed to her network of Nazi sympathizers because he was planning to write a book about the war: "Kent protested that he had not had a fair trial. He said he had no felonious intent, therefore he could not have feloniously obtained or communicated the documents." Unlike the trials that were conducted behind closed doors, the sentencing of both Wolkoff and Kent was open to the public.

The infamous Russian Tea Rooms, the location for so many of Wolkoff and Kent's seditious dealings, was closed soon after their sentencing. Admiral Wolkoff had arranged through his solicitors to file for bankruptcy so that whatever funds remained could be kept for his daughter when she left prison. Admiral Wolkoff's wife found work as the manager of a British government canteen and the former cook found employment in the household of "a very prominent person," whose identity was not disclosed. It was also noted that although the cook was Russian by birth, she spent her childhood in Germany and considers herself to be German.

There was still media interest in Anna Wolkoff, even in prison. Her sister, Kyra Wolkoff, was interviewed for an article in the *Daily Herald* titled "Spy Gives Dress Tips." Wolkoff's sister commented that Wolkoff had not appealed her sentence: "Anna has settled down and made up her mind to serve her sentence with a good grace. She is not moping because she is allowed to sew." When she was released from prison in June 1947, Wolkoff was assessed as "still likely to be a danger to the community." Having been stripped of her British citizenship, she lived mostly abroad until her death in a car accident in Spain in August 1973, aged seventy-one.

Kent was released from prison and immediately deported back to America, with a departure delay of two months due to a dock strike. His arrival on the SS *Silveroak*, which docked in Hoboken, New Jersey, in December 1945, was attended by fifty journalists. However, interest in Kent died down fairly quickly. He married and went on a yearlong cruise with his new wife. When he returned, he received word that the FBI wanted to question him in connection to a Nazi spy ring led by Kurt Jahnke. Intelligence from that investigation suggested that Kent was spying for the Russians and not for the Nazis. This line of inquiry was soon abandoned and Kent led a prosperous life on the East Coast as the owner of car dealerships, until the business went bankrupt in the mid-1960s.

Tried separately under Defence Act 18B, Sir Archibald Ramsay served four years in Brixton Prison. A "special source" reported to Maxwell Knight at MI5 that he selected the same solicitors who defended Wolkoff and that they waived their fees when they took his case. Ramsay's legal team filed a petition of poor treatment in prison that came to nothing. Word had traveled back to Ramsay that conditions for the Mosleys at the Royal Holloway Prison were considerably better. Mosley had a butler, a former petty thief, and Diana was allowed to wear her furs in her cell. When asked about whether members would follow Sir Oswald Mosley in the event of a "right wing revolution breaking out in this country," Ramsay responded "certainly not, before such a situation arises I shall be in touch with all the members and you will then be told who is to be your leader."

It was also requested that "when referring to Captain Ramsay in the future, he should be called "Freeman," a name that he had started to use when he was interned with his wife at Royal Holloway Prison. Perhaps coincidentally, Freeman was also the middle name of Unity Freeman Mitford, a red flag in the missing months of her intelligence file following her repatriation to England in January 1940.

CHAPTER 8

HITLER'S GIRL

By the early summer of 1939, war with Germany seemed inevitable. British nationals were told by the British government to leave Germany as tensions escalated and a larger-scale confrontation was on the horizon. While most foreigners left Munich on the government's advice, Unity Mitford made a point of returning to her adopted country with much fanfare, as breathlessly covered by the British tabloids. Departing from Dover on July 8, 1939, the Metropolitan Police searched her bags and noted that "the wooden cases taken to Germany by this woman were filled with tea and jam, as well as books, suggesting that she intends to remain on the continent for a lengthy period." By August 1939, even journalists hoping to cover the imminent outbreak of war had left. Unity, however, was determined to remain in Munich.

As early as spring 1938, Hitler had warned Unity to leave Germany. In an addendum to an inter-consular cable marked "good riddance," Ambassador Sir Nevile Henderson noted that Unity "saw Herr Hitler privately after his Nuremberg speech, when he advised her to leave Germany at once and to tell all her English friends to do the same as he felt certain there would be war. Miss Mitford is staying in Burgenland and did not wish to be warned to go, and if war broke out she would of course stay in Germany."

On the morning of September 3, 1939, the same day that Britain declared war on Germany, Unity Mitford walked to the Englischer Garten in Munich with a small pearl-handled pistol and supposedly fired a shot into her right temple. While some have speculated that the pistol was a gift from Adolf Hitler, it is more likely that Unity purchased it herself on his advice that a gun would be useful to protect herself given their close, and likely, romantic association. Just two weeks prior, after tea with Unity in Munich, Joseph Kennedy Jr., son of the American ambassador to England, wrote to his father that "she (Unity) is the most fervent Nazi imaginable, and is probably in love with Hitler." Hitler had warned Unity that if war broke out, his allegiance must be to his country, and not to her. If this was, indeed, a suicide attempt, Unity would be the third woman in Hitler's close circle to try to take her life. The führer's favorite astrologer had also predicted that Hitler would die by his own hand, which Hitler did not find ominous. He apparently believed that exceptional people must make exceptional sacrifices.

What was the daughter of an esteemed member of the House

of Lords doing in Munich after most British nationals had been evacuated on the eve of the Second World War? What happened that morning in Munich and what role did the British government play in covering up the events that followed?

In the spring of 1939, months before the dramatic episode in Munich's Englischer Garten, Unity abruptly moved out of the apartment of her friend Erna Hanfstaengl after denouncing Erna's brother Putzi, Hitler's close confidant and personal press secretary in the early days of the Reich. In addition to his official role, Putzi still frequently entertained the führer with German adaptations of fight songs he had learned at the Hasty Pudding Club as an undergraduate at Harvard. Putzi was, at one time, among Unity's closest friends and likely admirer, although he is perhaps more recently recognized for his romance with Martha Dodd, the controversial daughter of William Dodd, the American ambassador to Berlin from 1933 to 1937.

In those early days, Putzi had given Hitler entrée to social circles of potential supporters in Germany and England and connections to the United States. It was Putzi who first dangled the idea to Unity and Diana of coming to Germany to meet Hitler. But after he made an offhand remark praising the dynamic struggle in Franco's Spain over the de facto assumption of power by the Nazi Party in Germany, Unity questioned his loyalty to Hitler. That Hitler so quickly purged one of his most trusted advisers underlines Unity's influence.

As punishment for this supposedly disloyal statement, Hitler devised a sham mission over Spain with false orders to drop Putzi by parachute behind Spanish enemy lines so that he could experience the Spanish Civil War he had praised firsthand. However, instead of flying to Spain, Hitler instructed the pilot to fly in circles over Germany in order to disorient him and eventually land in Leipzig. Hanfstaengl was so alarmed by Hitler's deception that he left Germany immediately, was interned briefly as a war criminal in England, and was later brought to the United States by President Roosevelt to participate in the S-Project. Hanfstaengl is believed to have debriefed Roosevelt on as many as four hundred Nazi officers as well as provide information to the Allies about Hitler's personal life. Among those Ernst Hanfstaengl reported on to Roosevelt was Unity Mitford and her relationship with Hitler.

Hitler's treachery toward Hanfstaengl did not, however, frighten Unity. In fact, it made her bolder, although it did leave her without a place to live. Hitler responded so quickly to Unity's accusations against Putzi that she had to ask her old friend and likely former lover, the Hungarian count János Almásy, to pack her bags at Erna Hanfstaengl's apartment and move them to a hotel. Hitler ordered a young assistant, Gauleiter Wagner, to assist Unity in finding her own more suitable lodgings, where she would live alone for the first time since arriving in Germany nearly five years before.

Wagner showed Unity several apartments, all of which had been requisitioned from Jewish families after Kristallnacht in November 1938. She selected one on the Agnessestrasse near the Law Faculty of the Ludwig Maximilian University and a few blocks from the Englischer Garten in Munich. In a letter to her sister Diana, Unity wrote that the apartment had been vacated by "a young Jewish couple going abroad." She also noted the tearful couple's reaction when she decided to take the apartment. An avowed anti-Semite and fervent Nazi ideologue, it is difficult to imagine that Unity did not understand the real circumstances behind the apartment's procurement. Or more likely, she did not care.

On September 1, 1939, two days before England declared war against Germany, Unity gathered some documents in an envelope and delivered them to Gauleiter Wagner, then quickly departed. Two days later, she was supposedly found with a bullet wound to her temple on a park bench, not far from her apartment. The park bench where Unity was found was within a few hundred yards of a Nazi air district command post. Corporal Wincenty was immediately dispatched from his station and helped to secure the area while a passing tourist, Emil Knoblauch, witnessed Unity depart in an ambulance. She was taken to the Chirurgische Universitats-Klinik initially unidentified. The hospital staff who admitted her noted that she was not gravely wounded.

The circumstances surrounding what happened to Unity that morning in the Englischer Garten were ambiguous from the start. Unity claimed that she had been shot and later wrote to her sister Jessica that she had been "shot in the head . . . that paralyzed my right arm and right leg." But it was never clear whether a shot had actually been fired and, if so, by whom. Rumors circulated that rather than a suicide attempt, Unity was shot by members of Hitler's inner circle who sought to put an end to her influence over Hitler. The French journalist Geneviève Tabouis, German correspondent for the Marseilles newspaper *La Gironde* and the *Paris l'Oeuvre*, reported that Unity was shot by Gruppenführer August Scharenbach at Himmler's behest while walking in the Englischer Garten. Another journalist wrote that Unity "was sitting in the bedroom of her Munich hotel apartment recovering from an illness when Gestapo officers entered her room and a shot was heard." This theory was alluded to by Unity's uncle Jack Mitford who docked in New York the following February and remarked that Unity's bullet wound was not self-inflicted, although she had no memory of what had occurred.

There were other stories circulating in Germany about what happened to Unity that morning. Prince Nicholas Orloff, a Russian émigré and popular radio announcer for Berlin Radio, claimed that two sources who were close personal friends of Hitler separately declared that rather than a suicide attempt, Unity had argued with Hitler and taken an overdose of the barbiturate Veronal.

News of Unity's condition did not reach her parents in a timely way, which lends ambiguity to both what actually happened and the gravity of her health crisis. Classified files record a message from Unity's father, Lord Redesdale, to the British vice consul in Berlin on September 5, 1939, two days after the purported event and the outbreak of war with Britain. Redesdale requested that the vice consul phone him back at their London home at Kensington 5060. In a further conversation with the vice consul, Redesdale recounts that "he has begged her to come home but without success so far." Redesdale also requested that Unity's name be placed on a list of British citizens requiring assistance for repatriation by the American embassy in Berlin. A report to the Foreign Office on October 6, 1939, noted that "Lord Redesdale's daughter Unity, who is an intimate friend of Hitler, is in the meantime in Germany and is said to have changed, or to be about to change, her nationality to German." All of these exchanges occurred without mention of her hospitalization.

It is not until early November 1939, when Lady Redesdale requested information about Unity at the American embassy, that she was told that "when last heard of" Unity was in St. Joseph Hospital in Munich. She left a different phone number, Kensington 6476, in case any further information was received. In a letter to Lady Redesdale addressed to Old Mill Cottage in High Wycombe from the British Foreign Office, dated a few days after her inquiry, it was relayed that the American embassy had alerted the British embassy confirmation that "your daughter Miss Unity Mitford was in the Surgical Hospital at Munich and was well on the road to recovery. According to the information received by the Embassy, your daughter had attempted to do away with herself on September 3rd." The letter promised to keep Lady Redesdale updated with details about getting Unity sent back to England. The American ambassador sent a similar cable to the British secretary of state for foreign affairs, adding that Unity had not contacted any American officials. Lady Redesdale responded in a letter that "it is such a great relief to know that she is getting on well." Guy

Liddell, MI5's director of counterespionage, noted in his diary that "Lord Redesdale's daughter Unity has shot herself, having fallen in love with Hitler."

In late November 1939, an exchange of letters between British embassy personnel includes instructions that "I would very much like you to write to Unity's parents [at] 26 Rutland Gate and tell them that we visited her about 3 weeks ago in Munich and found her surrounded with every luxury and kindness possible. She has a lovely room and hopes to leave for home when the doctor allows it which should not be long now. A friend wrote two days ago to say she was getting on marvelously."

Unlike Unity's family, Hitler had been quickly notified of Unity's hospitalization, even amid the chaos of the German invasion of Poland. He personally ordered that she receive the finest medical care at his own expense. Albert Speer remarked how notable Hitler's attentions toward Unity were given the strategic complexity of the Polish invasion and subsequent Allied declarations of war. Hitler even asked Eva Braun, despite her own suicide attempt years earlier, when Unity first attracted Hitler's attention, to buy Unity whatever personal effects she needed during her hospital stay.

This was not the first time that Hitler had undertaken Unity's medical bills, many of which are still filed at the Bundesarchiv in Koblenz. Unity had fallen ill with pneumonia the previous summer, in August 1938, at the annual Bayreuth Festival, which she was planning to attend as Hitler's guest. Unity wrote in a letter to her family that when she became ill, she withdrew to her hotel room to avoid infecting Hitler as well. During her convalescence, Hitler sent her flowers, paid all of her medical bills, and even sent his own personal physician, Dr. Theodor Morell, to treat her with "ten or fifteen injections in a day," the same regime with which he treated Hitler. Hitler also called frequently from Berlin to check on her and "sent autographed pictures of himself for (Unity) to give to the nurses."

Although another Mitford daughter, Diana, also had ties to the Nazi Party and had been personally entertained by Hitler, Unity's father, Lord Redesdale, was concerned that Hitler's attention toward Unity in particular might suggest impropriety. As a result, Redesdale insisted on repaying all of Unity's medical bills to avoid further gossip. Lord Redesdale had only recently denied speculation about Unity's relationship with the führer to the *Sunday Pictorial*, insisting that "there is not nor has there ever been, any question of an engagement between my daughter and Herr Hitler. The Fuhrer lives only for his country and has no time for marriage."

While Hitler had once again stepped in to care for Unity, the invasion of Poland made circumstances more complicated. Detained at the front, Hitler called on Count János Almásy to watch over her as his proxy during this period, ironic on several accounts. Almásy, whose brother, László, has gained recent fame as *The English Patient*, was married to an invalid wife, Marie, and flew the swastika flag over the highest turret of his Schloss Bernstein in the remote Austrian Burgenwald. Almásy hosted shooting weekends and parties where both Hitler and Unity were frequent guests. Unity and Almásy were also rumored to have been lovers.

Countess Gaby Bentinck, whose family castle was located nearby, became a confidante of Unity's during this period. She described Almásy as "an astrologer and necromancer in the Wallenstein tradition, rather sinister-looking but invited everywhere, dashingly Nazi and Unity's bosom friend." Almásy was whispered to be Unity's "High Priest," spending hours with her in his library of occult texts where he supposedly initiated her into the realm of "gaspers" and the cult of Hitler's mysticism. Almásy explained to his eager pupil that Hitler's most ardent followers were preordained to die for the Nazi cause, most especially a young woman named Valkyrie, born in the town of Swastika, and now sworn to both Hitler and the Nazi cause.

The rumored intimate relationship between Almásy and Unity, which lasted just a few months, also presented another point of speculation about Unity's injury. It has been suggested that János Almásy might have made an attempt on Unity's life as retribution for her threats of blackmail on any number of accounts: his interest in the occult, which predicted that Hitler would eventually commit suicide as affirmation of his allegiance to the Nationalist Socialist cause; his bisexuality; or, more likely, his fear of Hitler's retribution for his intimate relationship with Unity. Hitler, however, was not deterred by any of these details and called on Almásy to keep a vigil by Unity's bedside and ensure that she received the best medical care available.

Despite the demands of the military incursion into Poland, Hitler personally arranged to store the furniture in Unity's Munich apartment and sent her diaries to Count Almásy for safekeeping.

Unity's diaries were carefully guarded by Almásy until the 1950s, when they were eventually turned over to her family. Unity's sister Deborah, later Duchess of Devonshire, was their most recent and final custodian. During her lifetime, Deborah prevented the circulation of Unity's diaries, and following her death in 2014, all access to both Unity and Deborah's papers, held at Chatsworth House in England, has been closed indefinitely.

In a hospital room filled with flowers from Ribbentrop, Goebbels, and Hitler, there are few first-person accounts of either Unity's condition or her prognosis. Hitler made a trip from the Polish front to Unity's hospital bedside on November 8, 1939, a visit that is noted in Martin Bormann's diary. Professor Schaub, the assistant to Professor Magnus, Unity's head doctor, described Hitler's visit: "he was deeply shaken by the fearful alteration to the beautiful, lively girl. Unity lay thoroughly apathetic and lamed in bed and took notice neither of the visitors whom she barely recognized, nor of the flowers they had brought." The attending physician, Dr. Helmut Reiser, noted that while Unity would not talk to the staff, she did speak when Hitler came to visit her, asking to return to England.

Not only did Hitler leave the Polish front to visit Unity, but this visit nearly cost him his life. He narrowly escaped an assassination attempt. Following a speech in the Bürgerbräukeller that ended at 9:07 P.M., Hitler boarded a train at 9:30 P.M. for Nuremberg. It was on the train that news of a bombing at the Bürgerbräukeller, approximately eight minutes after Hitler's departure, was received. A dynamite explosion had destroyed the Munich Bürgerbräukeller, leaving eight dead and sixty wounded.

Following Unity's request to Hitler to return to England, Dr. Reiser wrote that "the clinic was informed by the Bavarian Interministerium that contact had been made through diplomatic channels in Switzerland, and that Unity was to be taken there where English doctors would meet her for the journey back to Britain. A special carriage from the railways was to be prepared and a Sister of Mercy and a doctor (Reiser) were to accompany her." Under Almásy's watchful eye and with Hitler paying her way, Unity was transferred from the Munich Clinic to another hospital in Berne, Switzerland, in preparation for her return to England.

At Hitler's request, Almásy contacted the Interministerium for permission to accompany Unity and Dr. Reiser by medical ambulance from Munich to Berne. Reiser remembered that

we had this large railway carriage to ourselves, with a bed in it specially arranged for transporting the wounded. The sister sat beside her (Unity). The Count and I went to the dining car on the train, for he liked his food. I had the money from the Ministry of the Interior to pay expenses for myself and the Count at a hotel. An ambulance was waiting, Unity was transferred into it and went on to the clinic of Professor Matti. I think we had to wait until the English doctor came. But he didn't arrive and as I had personally handed over to Professor Matti all the medical documents, my mission was over. . . . Next day I visited Unity and Professor Matti asked, "What's all this about her not speaking? She has been singing Hitler's praises." . . . I waited until 24 December. I never saw the mother. I left. The Count visited me on his way back through Munich and told me that at least a week more had passed, and he had spent all the money, and had to ask for more at the embassy in Berne. We went to the Interministerium together and were told that it would be all right.

By Christmas Eve, an official at the British embassy in Berne, H. H. Hindmarsh, added a brief note to Unity's file that "Lord Redesdale told Unity that her mother and Stubby (Deborah) were coming out to her as soon as Christmas was over and they were able to get money for the journey. Unity said that she was keeping well though the journey to Berne had tired her. The suggestion was made that Unity should eventually go to live on 'her' island to recuperate. The nurse also joined in the conversation." The suggestion that an aristocratic family who traveled frequently to visit Unity in Germany as a stopover for extended European vacations, usually entertained by Hitler, might be suddenly lacking in funds to evacuate her from Switzerland is ironic. Unity's return did not seem to be pressing.

That same Christmas Eve 1939, MI5 transcribed a conversation between Lord Redesdale, Unity's father, and Count Almásy. In this conversation, Almásy requests to meet with Lord Redesdale at the Hotel Schweizer Hof, where he was staying. When Redesdale tells Almásy that he is not coming, but his wife and daughter are, Almásy responded, "No, no. I'd much rather you come. It is you that I want to see. Please, please try to come—and before the 27th." It is unclear what Almásy's rush was to return to Germany, or why he insisted that Lord Redesdale himself travel to Switzerland.

In the end, it was Lady Redesdale and her daughter Deborah who brought Unity back to England. With Europe at war, travel was difficult. While there is no record of their arrival in Switzerland, Lady Redesdale had phoned her husband at the Marlborough Club in London on New Year's Eve 1939 from a hospital in Berne about their planned return. As recorded by MI5 and transcribed in Unity's classified intelligence file, they discussed that the doctor taking care of their daughter had insisted on an expensive ambulance train to transport Unity from Paris to Calais. Lady Redesdale also asked that a stretcher be available at Folkestone to take their daughter off the boat. Lord and Lady Redesdale eventually spent sixteen hundred pounds for first-class rail passage from Berne to Calais. They also hired a guard van to follow their train.

• • •

WEDNESDAY, JANUARY 3, 1940. The muddy waters of Folkestone Harbor met the gray predawn sky as a steamship made its way to dock, the noise of its engines dulled by an icy drizzle marking the start of one of the coldest winters on record. The ship had arrived from Calais, one of the last to leave before the harbor was officially closed and repurposed for military use. The steamship was filled with weary travelers fleeing the Continent as Hitler continued to advance toward Denmark, Norway, France, and Belgium, all of which would be occupied by the early spring. The ship's most illustrious passengers, Unity Mitford, her sister Deborah, and their mother, Lady Redesdale, had traveled by special train from Paris to board the boat. This was the end of a journey of repatriation from Germany for Unity, who had traveled from Munich to Berne, Paris, and finally Calais, all but the last leg of the trip at Hitler's expense.

While the Redesdales planned Unity's repatriation, complaining about the expense and checking the logistics, word had leaked to the press about her arrival. Unity's relationship with Hitler and her widely publicized pro-Nazi views had earned her the name "Hitler's Girl" in the British press, which followed her frequent travel between Germany and England and reported on her fervent embrace of Nazi ideology with more fascination than criticism. The media descended on Folkestone Harbor as news had spread that Unity was returning from Germany under mysterious medical circumstances.

The jockeying between competing newspapers for the scoop on Unity's arrival was also recorded by the night duty officer of MI5. At 3.10 A.M. he noted a request by the *Daily Mail* for permission to "call their Amsterdam correspondent for details of the Unity Mitford story. CTC were unable to give a decision and rang up M.C.4. Officer Geoffrey allowed the call to be made while another duty officer stated that another paper had the news, and I asked him to confirm this and telephone what paper it was." Five minutes later, it was confirmed that the *Daily Express* had the scoop. At 3:25 A.M., the same duty officer reported that although the *Daily Mail* reporter complained bitterly about not being able to telephone Amsterdam from the Folkestone facilities provided to the press, he had filed a story that was then sent to Admiral Salmond for clearance.

With Unity's impending arrival at Folkestone, Sir Arthur Jeff, chief of field security police at Folkestone Harbor, received orders from MI5 to construct a tight security perimeter around the area as a member of Hitler's inner circle was being brought back to England. Rail traffic was diverted and highways were patrolled. A large police presence was put in place with all entrances to the port heavily guarded after a search of the entire perimeter of the harbor. Eventually, Lord Redesdale was admitted to the quay along with an ambulance he had rented from the Walton Private Ambulance Service. MI5 agents, unsure of which boat Unity was arriving on, maintained a constant vigil, updating each other regularly and exchanging plans to execute their orders to search Unity and her belongings and to have her examined by a doctor for a full assessment of her medical condition. The working assumption by the security services was that with England now at war with Germany, Unity stood to be accused of treason.

While a robust security presence was put in place, the Home Office intervened and quickly bypassed those measures. In fact, the roster of intelligence officers and local police on hand to meet the boat and assess Unity's security threat were barely notified of her arrival. Sir Arthur Jeff reported the last-minute change. "My final instructions received at 10:30am on Wednesday 3rd January, were that the party were to be treated as ordinary passengers. These were carried out except that out of consideration for the reported invalid, Miss Unity Mitford, the Chief Immigration Officer, acting on instructions from the Home Office to facilitate their passage through our 'control' himself carried out the routine interrogation, examination of passports etc. of the three ladies in their own cabin instead of in an exceedingly crowded and poky salon. . . . The Chief Immigration Officer informs me that this took him five minutes. . . . It is entirely untrue to say that I went into the cabin, questioned Unity etc. I never addressed a word to her: in fact, I only caught an accidental and momentary glimpse of her before the Chief Immigration Officer went into the cabin and shut the door."

The change in instructions about Unity's repatriation came from the highest ranks of the British government and were, in fact, the subject of a high-level government debate. Sir Guy Liddell, head of counterespionage for MI5, was posted to Folkestone Harbor for Unity's return and recounted the events of that day in his diary, code name Wallflower. On January 1, 1940, Liddell wrote, "[T]he question of Unity Mitford has been raised. It is reported that she has come out of Germany accompanied by some Hungarian and arrived in Switzerland, that she is in serious condition and that her mother and sister [Deborah] have gone over to fetch her." Liddell noted that he had phoned Sir Alexander Maxwell, the permanent undersecretary of state at the Home Office, to request that Unity be thoroughly searched if a medical officer deemed her to be fit enough, a request that Maxwell did not grant.

In addition to requests for a medical examination and search, Liddell also raised the prospect of internment on the grounds that

"Unity Mitford had been in close and intimate contact with the Fuhrer and his supporters for several years and was an ardent and open supporter of the Nazi regime. She had remained behind after the outbreak of war and her action came perilously near to high treason." Liddell added that if Unity had been a member of the general public and not a member of the British aristocracy, "the probability was that we should not be arguing the case, and that an Order under 18(b) would be made." The Home Office decided against pursuing internment charges because they had not taken any action against members of the Anglo-German Fellowship or the Link. "British citizens associated both groups who had visited Nazi Germany and maintained close ties to the Fuhrer's regime. . . . Her parents had been associated with the Anglo-German Fellowship and other kindred movements and had obviously supported her in her ideas about Hitler. . . . I still thought therefore that she should be served with an Order under 18(b) and that Lady Redesdale and [Deborah] Mitford should be searched discreetly but thoroughly."

On January 3, Liddell wrote that "Sir Alexander Maxwell rang up today about Unity Mitford and said that the home secretary was in entire agreement with his view and had decided that nothing should be done. Nobody was to be searched and no Order was to be Served. I reminded Maxwell once more that Unity was such a public figure there might be considerable criticism if she was left at large. He said that this aspect of the case had been considered."

With medical exams and searches officially abandoned by direct order from the home secretary, Sir John Anderson, Viscount Waverley—a cabinet member during the war—intervened to stop any search of the women as they left the ship. And so, Unity left the ship on a stretcher, wrapped in a blanket, never questioned or examined. As captured by a Pathé newsreel of the day's events, she is perfectly coiffed and her eyes alert as she covers her face and intermittently surveys the crowd, putting into question both the suggestion of a head injury and potential neurological impairment. The film reveals no visible mark where the bullet supposedly struck just three months before. In fact, not a hair is out of place. Unity was driven off the docks in the rented ambulance, which broke down several times and necessitated an overnight stay in a hotel, before reaching its destination at the Mitford family home in Swinbrook, Oxfordshire.

Following the controversy about her return to England and its subsequent documentation in the British press, Unity's MI5 file is dormant for several months after the final report about her arrival at Folkestone. There is not a single entry until the summer of 1940, nor is there any record of the medical examination mandated by the British security services. That examination eventually took place some nine months later, in September 1940, at the Radcliffe Hospital in Oxford, with a cursory report issued.

In the interim, the security services and Home Office fielded numerous complaints by the public about Unity being at large, seemingly in good health, and not prosecuted. A letter to Detective Inspector Ponsford at the Oxfordshire Constabulary confirmed that "Miss Unity Mitford is, and has been residing for some weeks with her mother Lady Redesdale, at Mill Cottage in the village of Swinbrook. . . . She has been seen motoring and shopping at

Burford, but always as a passenger in the car." Despite claims that she was incapacitated from the bullet wound and partially paralyzed from a car accident in Germany, the officer wrote, "I saw Miss Mitford and she appeared to have the normal use of her limbs." Another letter to the Oxfordshire police dated June 28, 1940, noted that "there is considerable feeling locally about her being at large. . . . If she is well enough to go shopping, she is well enough to be interned and ought to be in view of her past associations."

One postcard bearing a picturesque drawing of the Villa Carlotta, a famed botanical garden on Lake Como in Northern Italy and postmarked from London dated July 1940, was addressed to the home secretary, Sir John Anderson: "WHEN DO YOU PROPOSE TO INTERN UNITY MITFORD AND HER FATHER." Another addendum to her security file from the same time, likely from the MP for Epsom, Sir Archibald Southby, suggested that "if she is well enough to go about, she can receive all the necessary medical treatment in prison."

Herbert Morrison, a member of Parliament representing the newly formed Labour Party, challenged the decision not to intern Unity at a parliamentary debate, accusing the government of "taking steps to facilitate the return of a British subject who could be much better occupied in continuing the activities she was engaged in—assisting an enemy government." However, he did not maintain this position for long, claiming that "the circumstances under which she is living does not indicate that there is present any ground which makes it necessary in the interests of national security to exercise control over her." The response to that hasty change in opinion was captured by another MP, Mr.

Frankel, referring to Unity's "special privileges." He retorted that "this case makes people less convinced of the necessity for keeping hundreds of others in detention." A similar claim of privilege was advanced many years later by Kim Philby, one of the Cambridge Five. In a speech to East German Stasi officers, discovered in the Stasi archives after the fall of the Berlin Wall, Philby said that he was able to get away with treachery "because I was upper class."

Not only was Unity not interned, but she was well enough to attend the wedding of her youngest sister, Deborah Mitford, in London on April 20, 1941. The wedding of Deborah to Lord Andrew Cavendish, who soon after became the 11th Duke of Devonshire, was the event of a subdued London's wartime social season. Despite the union of two lofty aristocratic families, the headlines in the London newspapers heralded the event as the wedding of Unity Mitford's sister: "Unity's Sister Weds, Then Signs On." The *Daily Mirror*'s coverage claimed that "there can hardly have been a more elaborate gathering to celebrate a society wedding of what some might call front-line importance. One might have thought that the recent political history of Miss Unity Mitford and especially her Hitler-fawning record, might have kept away one or two of the most sensitive guests. Not on your life."

Wedding guests, including the Astors of Cliveden House and other prominent members of the right-wing clubs that the family frequented, were not deterred by speculation about whether Unity was carrying, as the *Daily Mirror* reported, "the special swastika badge that the Hun-gangster gave to her personally? Within a hundred yards of the place where this sparkling celebration took place the Fuhrer's evil handiwork gaped and yawned at the happy guests."

Unity and her mother entered the church by a side door to avoid the crowds. When she left for the wedding reception at the Mitford family's London residence, Unity "drove off with a woman companion. She shielded her face with a small blue *pochette* decorated with 'U.M.' in gilt letters." The *Sunday Pictorial* joined the *Daily Mirror* in its criticism of Unity, proclaiming that "Unity is a Flop" and that "The few people present were there to see the bride. NO ONE CARES ABOUT UNITY ANY MORE."

While public opinion about Unity collided with the group of like-minded wedding guests, her attendance confirmed her good health, again raising questions about her potential internment for treason. Declared "Well," Unity was the subject of a debate in the House of Commons, "Mr. Evelyn Walkden asked the home secretary whether his attention has been called to the fact that Miss Unity Mitford has now recovered from her illness; and whether he has considered the desirability of detaining her under Regulation 18B on account of her past close associations with Nazi leaders and her declared pro-Nazi views?" After a half-hearted debate in Parliament, once again no action was taken against Unity.

Speculation about both Unity's fitness and her whereabouts continued in Scotland, where she was rumored to be traveling to Inch Kenneth, a remote private island home in the Outer Hebrides, belonging to her family. A request for Unity to travel to Inch Kenneth had been previously turned down by the War Office soon after she returned to England. The suggestion that she was in Scotland was quickly dampened by the local Argyllshire Constabulary who confirmed that "there is no truth whatsoever in the allegation that Miss Unity Mitford was met by her father in Oban on Saturday, 20th June, but stated that her sister, Deborah—Lord Redesdale's youngest daughter—was met by her father at Oban that day and proceeded with him to Inch Kenneth."

By June 1941, inaction by the government contrasted sharply with renewed public outcry about Unity's free movement given her apparent good health. The local police interviewed Lord Redesdale, who confirmed that "she was genuinely in love with Hitler himself, though this love was mostly hero worship. . . . Hitler regarded Unity as a very sincere friend of his, but nothing more." Redesdale then repeated the story that she was "upset" when Germany declared war on Britain and went to the Englischer Garten in Munich with a small pistol she owned, "fired into the ground to see if her pistol was working then fired a shot into her right temple." He continued that "after she recovered consciousness, she was visited on numerous occasions by Hitler himself, and early in December 1939 Hitler sent a message through the Swiss government direct to Redesdale asking whether her parents desired her to be sent back to England." Redesdale also admitted that "there is no doubt that the health of his daughter has very greatly improved in the past year. She is physically active and strong. She still talks of her associations with Hitler."

Unity again corroborated the story about an attempted suicide. A neighbor in Swinbrook, Mrs. Phillimore, "a person of high social standing and worthy of confidence," told her hairdresser in

Burford that Unity had told her, "I have been a very silly girl and shot myself and the bullet is still in my brain."

The Oxfordshire police also interviewed Professor Cairns of the Radcliffe Infirmary, the "brain specialist" who was Unity's doctor. Cairns offered the opinion that "Unity Mitford is incapable of any intellectual activity that would materially assist enemy action in this country, and as a result of all enquiries I have made I entirely agree with this view. In the circumstances, therefore, no good purpose would be served either by interning this girl, or by placing her name on the Regional Suspect List." A confidential report from the Nuffield Department of Surgery dated June 5, 1941, nearly a year after her repatriation from Germany, concludes that Unity "has always been a person with a childish personality, and on September 4, 1939, in Munich, she shot herself in the head with a revolver, the bullet entered the right frontal region, traversed the brain from backwards and lodged in the left occipital lobe, where it still is. For a time after the injury, she had weakness of the right arm and leg, but this has now recovered. . . . She has considerable difficulty in concentrating and in initiating any activity, and she does not take any real interest in what is going on. There is some defect in reading and writing. At times she gets mildly depressed."

Discussions about Unity's recovery and possible imprisonment for treason were exacerbated by claims that she was having a relationship with a soldier, Pilot Officer John Sydney Andrews posted nearby to the small Cotswold town where she was living with her parents. According to a report submitted to MI5 on October 8, 1941, "Andrews was subsequently removed from Brize Norton to another station."

Constable Henry Brooks wrote that "this friendship caused a certain amount of gossip in the village and in fact Miss Mitford openly remarked amongst persons in the village that Pilot Officer Andrews was her lover, and although he was a married man with one child, he was going to marry her should he be able to obtain a divorce." Soon after, "Miss Mitford . . . informed persons in Swinbrook that her engagement with Pilot Officer Andrews had been broken off."

This public episode again brought more calls by the public for Unity to be interned, which again were turned down by the Home Office. While Unity was never interned, her visits to her sister Diana Mitford Mosley and husband Oswald Mosley at Royal Holloway Prison were carefully monitored. One prisoner remarked that "Miss Mitford was very excited and spoke very quickly and laughed a lot." In a report by the prison governor dated February 12, 1942, Unity said that she was thinking of changing her name to Rose Mary Smith to avoid future bad press. She claimed, "I know why they won't put me in prison. It is because Winston Churchill (a distant cousin) knows that I tried to commit suicide." She added to Diana, "I wish that they would put me in prison. I should love to be here with you." According to the prison governor's report, "Sir Oswald said she is so much improved and in fact if no one had known how very interesting and clever she was before, we would take her for quite a normal girl, and he thinks she is gradually growing up and will be quite her old self again."

After the flurry of visits to the Royal Holloway Prison, there does not seem to be much public interest in Unity once the hardship of war was brought close to home with food shortages and the nightly bombings of the Blitz. Unity was sent by her parents to live at the home of a local vicar in Warwickshire. The vicar slept in his dressing room while his wife slept with Unity. According to their daughter's recollection, all of the doors and windows were locked at night.

In August 1944, Unity applied for permission to reside in Inch Kenneth, the remote Scottish island that belonged to her family. The *Telegraph* reported that the home secretary granted approval "after considering special reports from the military intelligence and Special Branch at Scotland Yard," removing any consideration that the area was a sensitive military zone. "It does not seem necessary for HO (Home Office) to take any action with regard to her . . . the correspondent within appears to be alarmed lest plans for a third world war are being hatched on this island."

Unity lived an isolated life on the island of Inch Kenneth. She was suddenly rushed to the Scottish mainland hospital in Oban, where in May 1948 she died at the age of thirty-three of meningitis, an infection attributed to the bullet lodged in her brain. She was buried in the family plot in Swinbrook, with an epitaph on the simple headstone: SAY NOT THE STRUGGLE NAUGHT AVAILETH.

Among Unity's effects, given by her mother as a memento to a villager in Inch Kenneth, was a signed photo of Hitler from their time together. On the back of the photo, translated by Unity, was the inscription, YOU ARE BEAUTIFUL. YOU ARE MY EVERYTHING. MY HEART WILL ALWAYS BELONG TO YOU.

AFTERWORD: HITLER'S CHILD?

Unity's recently declassified MI5 file reveals how closely she was followed by the British intelligence service. While MI5 documented her proximity to Hitler, listened to her conversations, and monitored her views, neither the English government nor intelligence agencies capitalized on her access to advance their own cause or even their understanding of Hitler. In an era known for renowned double cross spying missions, Unity was never approached, recruited, or even considered by the British government as someone worth investigating. This is especially remarkable as she was the only British citizen known to have unfettered access to Hitler, who was quickly becoming a formidable enemy. The circumstances around Unity's repatriation to England raise more questions than her MI5's observers answer about who was watching Unity and to what end.

While frequent references are made of MI5's agreement that Unity Mitford should not be interned for treason, Guy Liddell did not believe that Unity had shot herself. Liddell was the senior MI5 officer tasked with meeting Unity at Folkestone Harbor when she returned from Nazi Germany. He wrote in his diary that he did not believe she had suffered a bullet wound to her temple. In an entry dated the day after her arrival, January 2, 1940, Liddell speculated that her exit from the ship on a stretcher was more likely a diversion to avoid further publicity and speculation about her condition.

When he pursued the idea of searching her and undertaking the agreed-upon medical examination before she departed the ship, the undersecretary at the Home Office, Alexander Maxwell, "was very reluctant to agree to this and thought that we should possibly make ourselves ridiculous." When Liddell tried to see Sir John Anderson, the home secretary, to raise a red flag about the omission in procedure, he had left the scene. Several days later, an entry in a report received by the SCO (security control officers) from an MI5 staff member who had interviewed Unity Mitford upon her arrival confirmed that there was no sign of a bullet wound: "we had no evidence to support the press allegations that she was in a serious state of health and it might well be that she was brought in on a stretcher in order to avoid publicity and unpleasantness to her family." This bombshell declaration, which indicates a cover-up at the highest level of the British government, was ignored, classified, and likely forgotten for over sixty-five years.

Besides his candid diary entries, did Liddell mention his findings to anyone at the intelligence agency? And if so, why was this information suppressed?

Of the many dead ends I encountered in researching this book, Unity's supposed love child was just one. The lively correspondence between the Mitford sisters also goes silent during the period when Unity returned from Germany to England. Even her intelligence file is dormant for several months after her return. Despite the consternation about their inability to question her or give her a physical exam to assess her exact medical condition, both overruled directly by the Home Office, British intelligence does not pick up Unity's trail again until nearly five months later. At that point, there are mostly acknowledgments of complaints written by locals about her living an ordinary country life, despite her treasonous conduct in Germany only months earlier. There is no record of the medical examination mandated by the British security services, which eventually did take place some nine months later, in September 1940, at the Radcliffe Hospital in Oxford.

Another explanation for the gaps in Unity's file surfaced in 2007. It offers a different account for Unity's departure from the ship on a stretcher, covered in a blanket, seemingly unharmed. Through reports by local villagers, we know that Unity spent a good part of the time that was unaccounted for in a small house in the village of Wigginton, in Oxfordshire, about seventeen miles from her family home. During the late 1930s and through the war, this house, Hillview Cottage, was used as a maternity home, a discrete location for the birth of local children, both legitimate and illegitimate. The cottage had two bedrooms, one bedroom for the nurse, Betty Norton, and one bedroom for the expectant mother.

According to Betty's niece, Val Hamm in December 2002, one

of those expectant mothers was Unity Mitford. Other locals claim that Unity gave birth to a son in the spring of 1940, a secret that was well known although carefully guarded in this small village, especially because it was widely accepted that Unity had given birth to Hitler's son.

The Oxfordshire birth registry confirms multiple births at Hillview Cottage during this period. According to Val Hamm in December 2002, Nurse Betty Norton confirmed that Unity Mitford had given birth at Hillview Cottage and "always said [the baby] was Hitler's."

While it would not have been unusual at that time for a birth certificate not to have been issued for an illegitimate child, the Oxfordshire birth registry does hold one tantalizing clue. It lists the birth of a boy in the small parish of Banbury, where Hillview Cottage was located, born to a mother named Freeman in the spring of 1940. Freeman was Unity's mother's maiden name, a common practice for naming illegitimate children during this period. Freeman was also an alias used by Archibald Ramsay while in prison. There is no corresponding record of this child's death, which suggests that not only was the possible birth of Hitler's son in England most likely covered up by the British government, but that he might well be alive today, a pensioner in the English countryside.

If the British government closed ranks and covered up Unity's pregnancy, then who was responsible and to what end? Was England, the cornerstone of democratic traditions, really as democratic as we have always believed? And just how close was this way of life to toppling? The secrets of this period lead to more nuanced conclusions—and questions.

When we look at the very familiar from the complicatedly similar perspective of today's world, we can now see that every small assault on democracy counted in pushing the momentum away from a shared sense of Western democratic values. We are reminded that each episode represents a small chip with a pointed chisel that was never questioned, explained away, or tolerated. If the forces of Fascist England that accelerated during the 1930s had prevailed, this period would have functioned as a prologue to ostensibly a very different outcome.

The rise of nationalism and populism, in both the United States and Europe today, bear many of the same hallmarks. Therefore, at what point does complacence become complicity, posing real risk to democratic norms that we take for granted? Will it require a similarly cataclysmic event like World War II in order to reassess and ensure democracy's survival?

There is much to learn from this period in the dusty archives, hidden away from view for so many years, and the stakes are high.

ACKNOWLEDGMENTS

It has been a privilege to work with my editor, Gail Winston, at HarperCollins. I have a fond memory of speaking with Gail about the proposal from Paris on a hot summer day. She was enthusiastic about this book from its earliest incarnation and made countless improvements to the manuscript. I also want to acknowledge the able support of Hayley Salmon as well as the rest of the team at HarperCollins. This book was written during challenging times for a researcher. The staff at the National Archives at Kew and the Beinecke Library at Yale both deserve mention. My students and colleagues at Yale and the London School of Economics are always a source of inspiration. Special thanks to my agent and friend, Elizabeth Sheinkman, who believed in this book from our first conversations and made sure it landed well. I am also lucky to have a family of trusted readers: Anna Sophia, Charlotte. and Paul.

NOTES

CHAPTER 1: GERMAN FASCISM CROSSES THE ENGLISH CHANNEL

18 "This is a family playing and momentarily referencing a gesture": BBC, "Queen Nazi Salute Film: Palace 'Disappointed' at Use," July 18, 2015.

18 "There is an extraordinary atmosphere of drama about a Mosley meeting": David Pryce-Jones, *Unity Mitford: A Quest* (London: Orion, 1995), p. 60.

18 "so much perfection argues rottenness somewhere": Beatrice Webb, *The Diary of Beatrice Webb,* ed. Norman and Jeanne MacKenzie, vol. 3 (Cambridge, MA: Harvard University Press, 1982–84), p. 418.

19 "which senile politicians have so long maintained on our public affairs": Jan Dalley, *Diana Mosley: A Biography of the Glamorous Mitford Sister Who Became Hitler's Friend and Married the Leader of Britain's Fascists* (New York: Alfred A. Knopf, 2000), p. 160.

19 "The average *Daily Mail* reader": *Spectator*, January 31, 1934.

20 "[A]lthough I loathe antisemitism, I do dislike Jews": Tony Kushner, *The Persistence of Prejudice: Antisemitism in British Society during the Second World War* (Manchester, UK: Manchester University Press, 1989), p. 2.

20 Lord Mount Temple, Lord Brabazon of Tara, and the Marquis of Clydesdale: Martin Pugh, *"Hurrah for the Blackshirts!": Fascists and Fascism between the War* (London: Random House, 2005), p. 5.

21   "made life worth living again": Dalley, *Diana Mosley: A Biography*, p. 159.

21   as well as a salary of £1 to £2 per week: Ibid., p. 160.

21   "'get rid of the Yids'": *Evening Standard*, November 2, 1936.

22   recently announced Nuremberg laws (1935) in Nazi Germany: Harry Defries, *Conservative Party Attitudes to Jews, 1900–1950* (London: Routledge, 2001), p. 121.

22   "of the unspeakable weapon of anti-Semitism": Jewish Labour Council Workers' Circle, *Sir Oswald Mosley and the Jews* (Aldgate, UK: Jewish Labour Council Workers' Circle, 1935), University of Warwick, Papers of Aaron Rapoport Rollin.

22   "I was embarrassed and slunk away": Pryce-Jones, *Unity Mitford*, p. 73.

23   "nails had pierced his head to the skull": Charlotte Mosley, ed., *Love from Nancy: The Letters of Nancy Mitford* (London: Hodder & Stoughton, 1993), p. 62.

23   rescinded his support after the violent clashes: A. K. Chesterton, *Portrait of a Leader: Oswald Mosley* (London: Sanctuary Press, 2019), p. 119.

23   "in the early days of the NSDAP (Nazi Party)": Letter from Unity to Sydney Redesdale. Unpublished, undated manuscript. Held at Chatsworth House Trust.

24   political role in their respective countries: Karina Urbach, *Go-betweens for Hitler* (Oxford: Oxford University Press, 2015), p. 167.

24   "whilst Bolshevism is the levelling of everything": Ibid., p. 171.

25   "the cancerous growth of Bolshevism": Churchill's speech delivered in Rome, January 20, 1927. Quoted in British National Archives HO 45/24893.

25   "accepted by millions of well-educated Germans": Urbach, *Go-betweens*, p. 185.

25   influence within the British royal family: Ibid., p. 187.

25   "race purity and fitness": Ibid.

26   where they both had homes: Charles Higham, *Mrs. Simpson: Secret Lives of the Duchess of Windsor* (London: Sidgwick & Jackson, 2004), p. 93.

26   "any such pressure on our daughter": Urbach, *Go-betweens*, p. 168.

27   with much fanfare at the university: Ibid., p. 183.

27 "assure me of Hitler's pacific intentions": John Julius Norwich, ed., *The Duff Cooper Diaries 1915–1951* (London: Phoenix, 2007), p. 218.

28 "diplomatic, political and industrial circles": Urbach, *Go-betweens*, p. 184.

28 "we will sit between two stools": Albert Speer, *Inside the Third Reich* (London: Weidenfeld, 1971), p. 71.

29 "in Germany and elsewhere": Adrian Phillips, *The King Who Had to Go: Edward VIII, Mrs. Simpson and the Hidden Politics of the Abdication Crisis* (London: Biteback Publishing, 2017).

29 "stretched out at them": British National Archives FO 371/18858.

CHAPTER 2: MORE DANGEROUS THAN THE BLACKSHIRTS

31 "the English popular press is dumb": Odette Keun, "Perfidious Albion," in *Tide and Time*, July 6, 1935.

32 "what we are going to do about them": Pugh, *"Hurrah for the Blackshirts!,"* p. 232.

33 allow them into their country: Alex Ross, "How American Racism Influenced Hitler," *New Yorker*, April 23, 2018.

34 lives to be taken more quickly: Ibid.

36 population of British citizenry: Pugh, *"Hurrah for the Blackshirts!,"* p. 231.

37 blond braids and "high spirits": Robin Saika, *The Red Book: The Membership List of the Right Club 1939* (London: Foxley Books, 2010), p. 26.

37 "A fur coat can wait": Ibid., p. 23.

38 "activities of Organized Jewry": Ibid., p. 136.

38 "Conservative Party of Jewish influence": Ibid., pp. 4–5.

38 "Jewish intrigue centered in New York": Ibid., p. 134.

39 "T'will serve to hang them yet": Ibid., p. 95.

39 a member of the Scottish Parliament: Ibid., p. 96.

40 "irreproachable sincerity and honesty": Ibid., p. 125.

40 of German propaganda: Ibid., p. 115.

41 "so vituperative, so vitriolic": Ibid.

41 "our Lord was a Jew": Ibid., p. 104.

42 "sometimes very well attended": Ibid., p. 43.

43 "I knew his uncle in India": British National Archives KV5/87.

44 "our own good Prime Minister": Saika, *The Red Book*, p. 27.

44 "without shedding one drop of blood": Dalley, *Diana Mosley: A Biography*, p. 245.

44 "the prospect of peace": *Policy of His Majesty's Government*, HC Deb 05 October 1938 volume 339 cc 337–454.

46 quick consolidation of power: Norman Rose, *The Cliveden Set: Portrait of an Exclusive Fraternity* (London: Random House, 2001), pp. 136–37.

46 "influence on the course of British policy": Ibid., p. 174.

47 as well as a few others: Ibid., p. 464.

47 "'Der Fuhrer, der Fuhrer'": Natalie Livingstone, *The Mistresses of Cliveden: Three Centuries of Scandal, Power and Intrigue* (London: Penguin Random House, 2015), p. 468.

48 "supports of German influence": Rose, *The Cliveden Set*.

49 "structure of British Democracy": Ibid., p. 5.

49 "German Foreign Policy": Livingstone, *The Mistresses of Cliveden*, p. 470.

49 "retire from public life": Rose, *The Cliveden Set*, p. 182.

50 "the crimes of Hitlerism": Ibid.

50 "senseless fable got into the headlines": Ibid., p. 181.

50 foreign policy closer to Germany: Ibid., p. 178.

51 "nothing but the thick stick": Ibid., p. 191.

51 "their own properties and privileges": Ibid., p. 192.

51 "a defeatist pampered group": Ibid., p. 194.

52 "I can't believe that has happened": Christopher Sykes, *Nancy: The Life of Lady Astor* (New York: Harper and Row, 1972), p. 414.

CHAPTER 3: A SHORT HISTORY OF THE LONG HISTORY OF BRITISH ANTI-SEMITISM

56 Britain was home to the third-largest Jewish community in Europe. This compares to Germany with about 525,000 Jews (0.75 percent of the total German population), Hungary with 445,000 (5.1 percent of the population), Czechoslovakia with 357,000 (2.4 percent), and Austria 191,000, most of whom lived in Vienna (2.8 percent). Source: United States Holocaust Memorial Museum.

56 Jews were "untrustworthy individuals": David Rosenberg, *Facing Up to Anti-Semitism: How Jews in Britain Countered the Threats of the 1930s* (London: JCARP Publications, 1985), pp. 14, 17.

56 synagogue was vandalized: Ibid., p. 5.

56 "Gentiles do not mix socially in numbers": Ibid., p. 17.

57 by the Gestapo—in Central London: Richard Dove, "Reviews: The Strange Case of Dora Fabian and Mathilde Wurm. By Charmian Brinson," *Journal of European Studies* 27, no. 4 (December 1, 1997): 474–75.

57 any statement about the case: Ibid.

57 "syndicate of dirty American Jews": Rosenberg, *Facing Up to Anti-Semitism,* p. 17.

57 "filth is swept right away": Ibid., p. 16.

57 imminent invasion by Germany: Michael Ffinch, *G. K. Chesterton* (San Francisco: Harper & Row, 1986), pp. 334–35.

58 "Any Jew Is Worth Two Englishmen": Pugh, *"Hurrah for the Blackshirts!,"* p. 230.

58 to benefit the British Red Cross: *Leaving Today: The Freuds in Exile 1938,* Exhibition at the Freud Museum London, July 18–September 30, 2018.

58 across a much larger population: David Litchfield, *Hitler's Valkyrie: The Uncensored Biography of Unity Mitford* (Stroud, UK: The History Press, 2013), p. 140.

59 "for all the evils of the day": Jewish Labour Council Workers' Circle, *Sir Oswald Mosley and the Jews.*

59 "unrest and political crisis": Ibid.

59 "pogroms in this country": Rosenberg, *Facing Up to Anti-Semitism,* p. 6.

60 "to help the German Jews": David Nasaw, *The Patriarch: The Remarkable Life and Turbulent Times of Joseph P. Kennedy* (New York: Penguin Press, 2012), p. 363.

60 "with this Jewish question": See Cabinet Files, British National Archives.

CHAPTER 4: THE DEBUTANTE NAZI

61 love of Wagnerian opera: Jonathan Guinness with Catherine Guinness, *The House of Mitford* (London: Phoenix, 2004), p. 213.

62 in the final days of the Reich: David Pryce-Jones, "'You Are Always Close to Me': Unity Mitford's Souvenirs of Hitler," *Spectator*, March 28, 2015.

62 since the Norman Conquest: Lord Montagu of Beaulieu, *More Equal than Others: The Changing Fortunes of the British and European Aristocracies* (New York: St. Martin's Press, 1970), p. 174.

62 "among the Christmas dolls": Anita Leslie, *The Gilt and the Gingerbread* (London: Hutchinson, 1981), p. 148.

62 "outstandingly beautiful tribe": Anita Leslie, *Cousin Randolph* (London: Hutchinson, 1985).

63 Hampstead League of Mercy: *Court Circular Times* (London), November 10, July 4, June 10, December 22, December 10, May 4, etc. [1932]

63 "flashing sham jewels": Jessica Mitford, *Hons and Rebels*, p. 66.

63 to liven up the party: Ibid.

63 girl of her social class: Leslie, *The Gilt and the Gingerbread*, p. 148.

64 her parents were eager to avoid: Nicholas Mosley, *Rules of the Game: Beyond the Pale: Memoirs of Sir Oswald Mosley and Family* (London: Dalkey Archive Press, 1991), pp. 252–60.

64 "first drawing-room Nazi": Dalley, *Diana Mosley: A Biography*, p. 153.

65 intimates and for several years: George S. Messersmith Papers, Special Collections, University of Delaware Library, *Report on Ernst Hanfstaengl*.

65 "during his leisure moments": Ibid.

66 "noises of the steam train": *Sunday Telegraph*, February 27, 2005.

66 eventually meet Unity and Diana: Erik Larson, *In the Garden of the Beasts: Love, Terror, and an American Family in Hitler's Berlin* (New York: Crown, 2011), p. 71. See Library of Congress, Martha Dodd Papers, Box 1, File 2, for calling cards Martha Dodd had received.

66 "permitted him to do anything": George S. Messersmith Papers.

66 "would not stand for that sort of thing": Ibid.

67 "flags in all the windows": Anne de Courcy, *Diana Mosley: Mitford Beauty, British Fascist, Hitler's Angel* (New York: William Morrow, 2003), p. 108.

67 ran through the crowd when he spoke: Ibid., p. 121.

68 "spirit which is moving Germany these days": Ibid., p. 158.

68 "I have ever been to in my life": Ibid., p. 107.

68 "I succeeded in that direction": Pryce-Jones, *Unity Mitford*, p. 78.

69 a decorated war hero, Captain Vincent: Ibid.

69 recommendation from Otto von Bismarck: Ernst Hanfstaengl,
Hitler: The Missing Years (New York: Arcade Publishing, 1994), p. 214.

69 "never let me go abroad again": Manuscript of unpublished
correspondence between Lady Redesdale and Unity Mitford,
undated. Held at Chatsworth House Trust.

69 "there was no one I would rather meet": Ibid.

69 at Nazi Headquarters in Berlin: Alexander Geppert, "Dear
Adolf!" Locating Love in Nazi Germany," in *New Dangerous Liaisons:
Discourses on Europe and Love in the Twentieth Century*, ed. L. Passerini and
A. Geppert (Oxford: Berghahn Books, 2008), pp. 190, 195–96.

70 "I can't possibly do without lipstick": de Courcy, *Diana Mosley:
Mitford Beauty*, p. 120.

70 extended family in the country: Ibid., p. 118.

71 "modest, but dignified baroness": Gioia Diliberto, *Debutante: The
Story of Brenda Frazier* (New York: Alfred A. Knopf, 1987), p. 78.

71 "the cook, was already old then": Pryce-Jones, *Unity Mitford*, p. 86.

71 "semi-romance" with Putzi Hanfstaengl: Charlotte Mosley, ed.,
The Mitfords: Letters between Six Sisters (New York: Harper, 2007), p. 43.

71 "she'd bothered to come out": Pryce-Jones, *Unity Mitford*.

72 "self-esteem was Perspex strong". Julia Boyd, *Travelers in the Third
Reich: The Rise of Fascism 1919–1945* (New York: Pegasus, 2018), e-book.

72 if she did not admit to liking Hitler: Ibid.

72 he would introduce her to Hitler: Ibid.

72 RESERVIERT FÜR DEN FÜHRER: Ibid.

73 "a special salute all to myself": Charlotte Mosley, ed., *Letters
between Six Sisters*, Letter from Unity to Diana, June 12, 1934.

73 the Night of the Long Knives: Litchfield, *Hitler's Valkyrie*, p. 127.

74 "The whole thing is so dreadful": Charlotte Mosely, *Letters between
Six Sisters*, Letter from Unity to Diana, July 1, 1934, pp. 47–48.

74 "Miss Hanfstaengl": Litchfield, *Hitler's Valkyrie*, p. 129.

74 "enormously fat": Charlotte Mosley, *Letters between Six Sisters*,
Letter from Unity to Diana, May 4, 1934.

75   "she was my guest": See Pryce-Jones, *Unity Mitford*, pp. 156–57.

75   "(she) was hoping to see Hitler": Litchfield, *Hitler's Valkyrie*, p. 139.

75   "the big meetings, they are thrilling": Unpublished, undated letter from Unity Mitford to Lady Redesdale. Held at Chatsworth House Trust.

76   "had their *Stammitsch* and came every day": Pryce-Jones, *Unity Mitford*, p. 103.

76   "one of the most thrilling moments of my life": Unpublished, undated letter from Unity Mitford to Lady Redesdale. Held at Chatsworth House Trust.

76   "vegetable cutlets": Ibid.

76   "were really all she had": Charlotte Mosley, *Letters between Six Sisters*, Letter from Unity to Diana, December 23, 1935, p. 68.

77   "And it paid off": Pryce-Jones, *Unity Mitford*, p. 86.

77   at the Osteria Bavaria: Otto Dietrich, *The Hitler I Knew*, trans. Richard and Clara Winston (London: Methuen, 1957), p. 188.

77   "go by her table on his way out": Dalley, *Diana Mosley: A Biography*, p. 119.

78   "thrilled to death of course": Pryce-Jones, *Unity Mitford*, p. 95.

78   "passing through Munich": Ibid.

78   "I did not like him": Unpublished, undated letter from Unity Mitford to Lady Redesdale. Held at Chatsworth House Trust.

78   "my little sunshine": Pryce-Jones, *Unity Mitford*, p. 147.

79   Unity was English and not German: Unpublished, undated letter from Unity Mitford to Lady Redesdale. Held at Chatsworth House Trust.

79   "she addresses him as *Mein Fuhrer*": Ibid.

79   "possibility of a love affair with her": Pryce-Jones, *Unity Mitford*, p. 96.

79   "the police knew it and never did a thing": Ibid., p. 147.

79   on account of the "international Jews": Guinness and Guinness, *The House of Mitford*, pp. 370–71.

80   "he speaks a lot in analogies": Letter to Lord Redesdale, February 19, 1935. Held at Chatsworth House Trust.

80   she wouldn't mind dying: Boyd, *Travelers in the Third Reich*, e-book.

80   "Sweet Uncle Wolf": de Courcy, *Diana Mosley: Mitford Beauty*, p. 151.

80 "Hero-worship": Pryce-Jones, *Unity Mitford*, p. 96.

80 "& talked a lot about all these Notes": Likely diplomatic notes that were exchanged when Hitler violated the military clause of the Treaty of Versailles.

81 "he has a very nice face": Charlotte Mosley, *Letters between Six Sisters*, pp. 55–56.

81 "full of praise for what they had seen": Mary S. Lovell, *The Sisters: The Saga of the Mitford Family*, 1st American ed. (New York: W. W. Norton, 2002).

82 "found him intelligent to talk to": Charlotte Mosley, *Letters between Six Sisters*, p. 56.

82 for the week he visited: Ibid., p. 57.

82 "She said what came into her head": Diana Mosley, *A Life of Contrasts: The Autobiography of Diana Mosley*, American ed. (New York: Time Books, 1977), p. 153.

83 "tea in the Fuhrer's home": Ernest Pope, *Sunday Dispatch*, January 25, 1942.

83 "lord of that National Socialist Party": Ernest Pope, *Sunday Dispatch*, January 25, 1942.

84 "he appreciated frankness in her": Pryce-Jones, *Unity Mitford*, pp. 96–97.

84 "knowing that rumours would spread": Ibid., p. 97.

84 "I have only to tell her": Ibid., p. 148.

84 "the person that everyone was interested in": de Courcy, *Diana Mosley: Mitford Beauty*, p. 152.

84 "that 'good for nothing Unity'": Litchfield, *Hitler's Valkyrie*, p. 130.

85 "I could write a funny book about them": Charlotte Mosley, *Letters between Six Sisters*, p. 61.

86 "A pleasant host": Litchfield, *Hitler's Valkyrie*, p. 158.

86 "unaware that we knew each other": Sir Oswald Mosley, *My Life*, 2nd ed. (New Rochelle, NY: Arlington House, 1972), p. 368.

86 "in no way neurotic": Ibid., pp. 365–66.

86 Friederike, later the Queen of Greece: Brigitte Hamann, *Winifred Wagner: A Life in the Heart of Hitler's Bayreuth*, trans. Alan Bance (New York: Harcourt, 2005), p. 240.

87 "it was the fashion to chatter on like that": Litchfield, *Hitler's Valkyrie*, p. 130.

87   "such a great personality at its head": "She Adores Hitler," *Sunday Express*, May 28, 1935.

87   "The New Age has come": Unpublished letter from Unity Mitford to Lady Redesdale, June 4, 1935. Held at Chatsworth House Trust.

88   "friendship between Germany and England will prevail": Pryce-Jones translation in *Unity Mitford*, pp. 114–15.

88   "she sees Herr Hitler very often in Munich": British National Archives FO 371/18859.

89   "getting faster and faster": Letter from Unity Mitford to Lady Redesdale, June 23, 1935. Held at Chatsworth House Trust.

89   as she left the stage: Ibid.

89   "Queen of the May this year at Hesselberg": Charlotte Mosley, *Letters between Six Sisters*, Letter from Nancy Mitford to Unity Mitford, June 29, 1935, p. 60.

89   "My heart will always belong to you": David Pryce-Jones, "Eva Braun's Rival: The British Socialite Who Loved Hitler," *National Review*, March 21, 2015.

90   "against my own conscience, and it doesn't": Unity Mitford to Lady Redesdale, undated manuscript. Held at Chatsworth House Trust.

90   "Germany has already completed victoriously": Harry Kessler, *Berlin in Lights: The Diaries of Count Harry Kessler, 1918–1937*, trans. Charles Kessler (New York: Grove Press, 2000), p. 474.

91   "Out with the Jews!": *Der Stürmer*, July 1935 issue, p. 6.

91   "The dagger is lovely": Letter from Unity Mitford to Lady Redesdale, undated. Held at Chatsworth House Trust.

91   "to get her name in the papers": *Jewish Chronicle*, August 2, 1935.

92   "lunch parties in his flat": Michael Burn, *Turned towards the Sun* (Wilby, Norwich, UK: Michael Russell Publishing, 2003).

92   "were greeted uproariously": Pryce-Jones, *Unity Mitford*, pp. 121–22.

93   "She was a bit of a suffragette": Ibid., p. 123.

93   "He is happy to be rid of them": Letter from Unity Mitford to Lady Redesdale, December 23, 1935. Held at Chatsworth House Trust.

93   "'sends you faithfully Adolf Hitler'": Ibid., December 25, 1935.

93 for permission to visit Dachau: Ibid., January 1936.

94 "in the roof of the temple": Letter from Unity Mitford to Diana Mitford, February 2, 1936. Held at Chatsworth House Trust.

CHAPTER 5: "SWEET DR. GOEBBELS"

95 "labor unions in a surprisingly short time?": Litchfield, *Hitler's Valkyrie*, p. 104.

96 "silly people at Nuremberg": British National Archives FO 371/18858.

96 "rehabilitate Germany in despite of England": John Heygate, *These Germans: An Estimate of Their Character Seen in Flashes from the Drama 1918–1939* (London: Hutchinson, 1940), p. 180.

97 "rather than be over-pernickety": Dalley, *Diana Mosley: A Biography*, p. 220.

97 "we are the poorer thereby": Ibid.

97 "It is simply not true": Ibid., pp. 221–22.

98 "our victory is certain": Pryce-Jones, *Unity Mitford*, p. 111.

98 the group's activities to MI5: Litchfield, *Hitler's Valkyrie*, p. 174.

99 "a tragedy was averted": *Anglo-German Review*, November 1936.

100 "don't you think": Charlotte Mosley, *Letters between Six Sisters*, Letter from Unity Mitford to Diana Mitford, February 8, 1936, p. 69.

100 "partners will be changed": Juliet Gardiner, *The Thirties: An Intimate History* (London: Harper Press, 2010), p. 429.

100 "friendship between the two Nordic countries": Ibid.

101 winning fourteen medals: Ibid., p. 431.

101 the release of a thousand pigeons: de Courcy, *Diana Mosley: Mitford Beauty*, p. 172.

101 "has won the masses": Gardiner, *The Thirties*, p. 430.

102 the influential British daily: Ibid.

102 "pro-German but pro-Nazi": Dalley, *Diana Mosley: A Biography*, p. 215.

102 "supposed interests of your adopted class": Ibid., p. 217.

102 private secretary to Prime Minister David Lloyd George: Ibid., p. 216.

103 "She cried and I was so sorry": de Courcy, *Diana Mosley: Mitford Beauty*, p. 172.

103 not in the main house: Ibid.

104 "without a trace of Teutonic flavor": Dalley, *Diana Mosley: A Biography*, p. 217.

104 "would have impressed the Romans": Ibid., p. 218.

104 "because everything had geklappt (worked)": Charlotte Mosley, *Letters between Six Sisters*, p. 76.

105 the arrest of eighty-three protesters: Dalley, *Diana Mosley: A Biography*, p. 223.

105 "I want to leave Bryan": Litchfield, *Hitler's Valkyrie*, p. 109.

106 an ankle-length black skirt: Dalley, *Diana Mosley: A Biography*, p. 174.

106 "and your sweetness too": Charlotte Mosley, *Letters between Six Sisters*, p. 77.

107 "for your success and happiness Magda Goebbels": Dalley, *Diana Mosley: A Biography*, p. 225.

107 "the best I ever heard him make": Ibid., p. 224.

107 "'newspapermen could squirt their poison'": Sefton Delmer, *Sunday Express*, October 11, 1936.

108 "you are still in the Fuhrer's train": Charlotte Mosley, *Letters between Six Sisters*, p. 77.

108 "we had a lot of jokes": Ibid., p. 78.

108 "Tell him that please!": "Who is King Here?" "Strictly Confidential" report was addressed "Only for the Führer and Party Member v. Ribbentrop."

CHAPTER 6: WHOSE SIDE?

110 avoid a war between the two countries: Federal Bureau of Investigation File Number HQ65-31113 and HQ94-46650. Released by the Freedom of Information and Privacy Acts. Undated.

111 "her conversation deft and amusing": Ibid., p. 39.

112 at the end of the week: British National Archives FO/954/10A.

112 "because of that stupid tactlessness": Testimony of Reichsmarshal Goering to Major Douglas Kelley, November 11, 1945, International Military Tribunal, *Trial of the Major War Criminals before the International Military Tribunal* (Nuremberg: Allied Control Authority of Germany, 1947), IX, 400.

113 "the last thing which any of us desire": British National Archives FO/054/10A.

113 "every drop of blood in my veins is German": Morton, *17 Carnations*, p. 57.

113 "for what reason I cannot imagine": Ibid., p. 63.

114 "interesting and significant": Ibid., p. 58.

114 "with which he was treated in Austria": British National Archives FO 954/33.

114 "His abdication was a severe loss for us": Speer, *Inside the Third Reich*, p. 72.

114 "formation of British foreign policy": Morton, *17 Carnations*, pp. 60–61.

115 "There is no genuine trait in him": Paul Schwarz, *Ribbentrop the Man* (New York: Julian Messner, 1943), pp. 294–95.

115 "that she was passing on to the Germans": Morton, *17 Carnations*, p. 158.

116 "cast an undesirable reflection upon their King": Ibid., p. 74.

116 "'in several of the larger cities'": British National Archives FO 954/33.

117 "would not make any speeches": Ibid.

117 "costs of the trip would be paid by Germany": Ibid.

117 "regarded as countenancing it": Ibid.

117 "as little as possible to do with them": Ibid.

118 "regarded as countenancing the visit": Ibid.

118 "in no way approves of the visit": Ibid.

118 "it is not possible to stop the Duke of Windsor going to Germany": British National Archives FO 954/33A-55.

119 "A great man!": Morton, *17 Carnations*, p. 132.

119 "'where they store the cold meat'": Ibid.

120 "with the same peculiar fire": Greg King, *The Duchess of Windsor: The Uncommon Life of Wallis Simpson* (London: Citadel, 2020), e-book.

120 "it was a soldier's salute": Morton, *17 Carnations*, p. 133.

121 a "semi fascist comeback in England": Ibid., p. 137.

121 "to marry her and maintain the throne": Ibid., p. 158.

121 "England's destiny after the war": Ibid., p. 163.

122 "the suggestion comes from Germany": British National Archives CAB-301-179_01.

123 "make England ready for peace": Morton, *17 Carnations*, p. 178.

123 "became very thoughtful": British National Archives
CAB-301-179_01.

124 "both to HM and the Government": Morton, *17 Carnations*, p. 174.

124 "to do away with him at first opportunity": British National
Archives CAB-301-179_01.

125 "as good as impossible": Ibid.

125 on behalf of Germany: Ibid.

126 "Spanish secret police will ensure a safe crossing": British
National Archives CAB-301-179_12.

126 "Duke and Duchess are prepared to return to Spain": Ibid.

126 "he would give an answer in 48 hours": Ibid.

126 "in accordance with their wishes": British National Archives
CAB-301-179_13.

127 "expressed admiration and sympathy for the Fuehrer": British
National Archives CAB-301-179_15.

127 "persuade the Duke and Duchess to remain in Europe": British
National Archives CAB-301-179_16.

127 "Security Police could offer no guarantees": Ibid.

128 "Shall any answer be sent?": British National Archives
CAB-301-179_18.

128 "was listening to suggestions that were disloyal": Ibid.

129 "there was no possible value to them": British National Archives
CAB-301-179_19.

129 "destroy all traces of these German intrigues": Ibid.

130 "and never saw him after 1937": Federal Bureau of Investigation
File Number HQ65-31113 and HQ94-46650. Released by the
Freedom of Information and Privacy Acts. Undated.

130 "he is virtually non compos mentis": Ibid.

131 ensure that the duchess had no communication with Ribbentrop:
Ibid.

131 "the transferring of messages through the clothes may be taking
place": Ibid.

131 "among the British aristocracy": Ibid.

132 "represented the number of times they had slept together": Ibid.

134 "it will cost the lives of millions of Americans": Ibid.

134 "there would be no question about his wife's being queen": Ibid.

135 during the Windsors' visit to Miami and Palm Beach: Ibid.

135 "praised the German government and the Hitler regime": Ibid.

136 "I'll fit in anywhere that I can be helpful": Ibid.

136 "how pleasant it was to renew my acquaintance with you": Ibid.

137 Anglo-German Naval Agreement (1935) with the Chamberlain government: See Avalon Project, Yale Law School.

137 "policy did not materialize": *Nuremberg Trial Proceedings*, vol. 9, March 16, 1946, Avalon Project, Yale Law School.

138 shared with President Truman and Marshal Stalin: British National Archives FO/954/32D.

CHAPTER 7: THE WOLKOFF AFFAIR

139 "It may be here (in the US) too": Op-Ed by Ambassador Joseph Kennedy, *Boston Sunday Globe*, November 10, 1940.

141 "outside the courtroom would see a certain witness": Staff reporter, *Daily Express*, October 24, 1940.

141 likely to be high-ranking MI5 agent Maxwell Knight: British Press Association, October 1940.

142 further raised the suspicions of British intelligence: British National Archives KV-2-841_1.

142 "they will not be able to understand": Ibid.

143 whatever she wished: Ibid.

144 "Hitler was undoubtedly a genius": British National Archives KV-2-842_6.

144 both MI5 and the police. British National Archives KV-2-841_1.

144 "membership of the Right Club was a secret matter": British National Archives KV-2-2_6.

145 "for the way it had rid itself of Jews": British National Archives KV-2-841_2.

145 "wears a gold-rimmed monocle": British National Archives KV-2-841_1.

145 he would spit in the faces of the English: Ibid.

145 Imperial Russian War Cabinet of 1914: Ignatieff was the father of former Canadian MP and Opposition Party leader Michael Ignatieff who now serves as the president of Central European University.

146 "until the outbreak of the war": British National Archives KV-2-842_6.

147 "in Harrington Gardens almost every night": British National Archives KV-2-841_4.

147 with whom she was in frequent contact: British National Archives KV-2-841_1.

147 "Jews were ruining the country": British National Archives KV-2-841_2.

147 the two women made omelets together: Ibid.

148 "England would not be able to stand up to Germany": Ibid.

148 "could be read with a magnifying glass": Ibid.

148 "strategic position of the Allies": British National Archives KVI-2-841_3.

149 "you will give us the stuff all the same": Paul Willetts, *Rendezvous at the Russian Tea Rooms: The Spy Hunter, the Fashion Designer and the Man from Moscow* (London: Constable, 2015), p. 324.

150 "subversive operations in Great Britain": British National Archives KV-2-842_6.

150 to Maxwell Knight of MI5 instead of Joyce: British National Archives KV-2-841_2.

151 with "pluck and initiative": Ibid.

151 "dark hair and complexion": British National Archives KV-2-841_1.

151 "looked like a cheap gangster": British National Archives KV-2-1698.

151 "obtaining 'intelligence' in this country": Ibid.

152 might be an accomplice: Ibid.

152 to the Embassy Club where they stayed until just after midnight: British National Archives KV-2-841_2.

152 "I made no further attempt to get in touch with Kent": Ibid.

153 "asked me to keep it for him": Ibid.

154 "where any unauthorized person might have access to it": British National Archives KV-2-841_1.

154 "removed to prison as a menace to that policy": Ibid.

154 disclosure of unauthorized information: British National Archives KV-2-841_4.

154 evidence of her political leanings: Ibid.

155 "my duty to make these facts known": British National Archives KV-2-841_1.

155 "the telegrams were very secret": Ibid.

155 "both the governments concerned": British National Archives KV-2-841_3.

156 "they would be of use to an enemy": British National Archives KV-2-841_1.

156 "only because she was in love with him": British National Archives KV-2-841_6.

156 that Kent was a spy—and suggest a love affair: Ibid.

156 "to the full extent of her power": British National Archives KV-2-841_2.

156 "without any specific object in mind": Ibid.

156 she spent an hour there alone: Ibid.

157 was cleared when each count was presented: British National Archives KV-2-841_4.

157 "used as a means of dissolving Parliament": Roosevelt spelled incorrectly in the text. British National Archives KV-2-841_6.

158 whose identity was not disclosed: Ibid.

158 considers herself to be German: Ibid.

158 "because she is allowed to sew": Ibid.

159 car accident in Spain in August 1973, aged seventy-one: Ibid.

160 "you will then be told who is to be your leader": British National Archives KV-2-841_4.

160 with his wife at Royal Holloway Prison: British National Archives KV-2-841_1.

CHAPTER 8: HITLER'S GIRL

161 "she intends to remain on the continent for a lengthy period": British National Archives KV-2882_64.

162 "she would of course stay in Germany": British National Archives KV2/882.

162 "probably in love with Hitler": August 22, 1939. Amanda Smith, ed., *Hostage to Fortune: The Letters of Joseph P. Kennedy* (New York: Penguin, 2001), 355–56.

163 American ambassador to Berlin from 1933 to 1937: See Larson, *In the Garden of the Beasts.*

163 Unity questioned his loyalty to Hitler: Litchfield, *Hitler's Valkyrie,* p. 196.

165 "a young Jewish couple going abroad": Lovell, *The Sisters*, p. 286.

165 witnessed Unity depart in an ambulance: Bayerische
Hauptstaatsarchiv, Munich, September 3, 1939 (Pol. Dir. Munchen,
10 117).

165 noted that she was not gravely wounded: Pryce-Jones, *Unity
Mitford*, p. 297.

166 "Gestapo officers entered her room and a shot was heard": Ibid.,
pp. 306–7.

166 she had no memory of what had occurred: Ibid., p. 307.

167 repatriation by the American embassy in Berlin: British National
Archives KV2/882 Ref 11410.

167 "about to change, her nationality to German": British National
Archives KV-2882_63.

167 getting Unity sent back to England: British National Archives
KV2/882 11410/11419/218.

167 "she is getting on well": British National Archives KV2/882
Ref 113.

168 "having fallen in love with Hitler": *The Guy Liddell Diaries*, Volume
1: *1939–1942 MI5's Director of Counter-Espionage in World War II*, ed.
Nigel West (New York: Routledge, 2005), p. 54.

168 "she was getting on marvelously": British National Archives
KV-2-882_61.

168 subsequent Allied declarations of war: Albert Speer, *Spandau: The
Secret Diaries* (New York: Macmillan, 1976), p. 331.

169 "to give to the nurses": *Times of London*, August 19, 1938.

170 "Unity's bosom friend": Litchfield, *Hitler's Valkyrie*, p. 130.

170 his intimate relationship with Unity: Ibid., p. 228.

170 to Count Almásy for safekeeping: Pryce-Jones, *Unity Mitford*, p. 301.

171 "the flowers they had brought": Ibid., p. 295.

171 "eight dead and sixty wounded": Ibid.

172 "doctor (Reiser) were to accompany her": Ibid., p. 298.

173 "told that it would be all right": Ibid., p. 299.

173 "The nurse also joined in the conversation": British National
Archives KV-2-882_60.

173 "and before the 27th": British National Archives KV-2-882_59.

174 hired a guard van to follow their train: Pryce-Jones, *Unity
Mitford*, p. 307.

175 "and telephone what paper it was": British National Archives KV-2-882_57.

176 sent to Admiral Salmond for clearance: British National Archives KV-882.

176 Walton Private Ambulance Service: Pryce-Jones, *Unity Mitford*, p. 308.

177 "went into the cabin and shut the door": British National Archives KV-882_47.

177 "her mother and sister [Deborah] have gone over to fetch her": *The Guy Liddell Diaries*, Volume 1, ed. West, p. 54.

177 a request that Maxwell did not grant: Ibid., p. 55.

178 "came perilously near to high treason": Ibid.

178 "should be searched discreetly but thoroughly": Ibid., p. 54.

178 "the case had been considered": Ibid.

180 "in view of her past associations": British National Archives KV-882, June 28, 1940.

180 "necessary medical treatment in prison": British National Archives KV-882.

180 "assisting an enemy government": British National Archives KV-882, January 24, 1940.

181 "keeping hundreds of others in detention": British National Archives KV-882, January 24, 1940.

181 "Unity's Sister Weds, Then Signs On": "Unity's Sister Weds, Then Signs On," *Sunday Dispatch*, April 20, 1941.

181 "Not on your life". *Daily Mirror*, April 22, 1941.

182 "yawned at the happy guests": Ibid.

182 "'U.M.' in gilt letters": *Daily Express*, April 20, 1941.

182 "NO ONE CARES ABOUT UNITY ANY MORE": "NO ONE CARES ABOUT UNITY ANY MORE," *Sunday Pictorial*, April 20, 1941.

182 "her declared pro-Nazi views?": British National Archives KV-882.

183 "sent back to England": Ibid.

183 "talks of her associations with Hitler": Ibid.

184 "the bullet is still in my brain": Ibid.

184 "she gets mildly depressed": Ibid.

185 "removed from Brize Norton to another station": Ibid.

185 "had been broken off": Ibid.

185 "spoke very quickly and laughed a lot": Ibid.

185 "I tried to commit suicide": Ibid.

185 "I should love to be here with you": Ibid.

185 "quite her old self again": Ibid.

186 MY HEART WILL ALWAYS BELONG TO YOU: David Pryce-Jones, "You Are Always Close to Me," *Spectator*, March 28, 2015.

AFTERWORD: HITLER'S CHILD?

188 "avoid publicity and unpleasantness to her family": British National Archives KV4/19.

190 "always said (the baby) was Hitler's": Martin Bright, "Unity Mitford and Hitler's Baby," *New Statesman*, September 27, 2015.

INDEX

ABOUT THE AUTHOR

LAUREN YOUNG is an academic and policy consultant specializing in security and defense issues. She is currently a lecturer at Yale University, where she teaches in the Department of Political Science and the Jackson School for Global Affairs. She lives in New York City.